WOLF LAND

WOLF LAND

The Lost Wolves of Landscape and Lore

ELIZABETH MARSHALL

PELAGIC PUBLISHING

First published in 2026 by
Pelagic Publishing
20–22 Wenlock Road
London N1 7GU, UK

www.pelagicpublishing.com

Wolf Land: The Lost Wolves of Landscape and Lore

https://doi.org/10.53061/TKEL9384

A CIP record for this book is available from the British Library

EU Authorised Representative: Easy Access System Europe –
Mustamae tee 50, 10621 Tallinn, Estonia, gpsr.requests@easproject.com

ISBN 978-1-78427-396-5 Hbk
ISBN 978-1-78427-397-2 ePub
ISBN 978-1-78427-398-9 PDF

Cover artwork by Rachel Hudson

Typeset in Minion Pro by S4Carlisle Publishing Services, Chennai, India

5 4 3 2 1

Printed and bound by Short Run Press Ltd, Exeter EX2 7LW

MIX
Paper | Supporting responsible forestry
FSC® C014540
FSC www.fsc.org

For the wolves who lived and died in Britain and Ireland, who deserved better than the fate they were dealt by human hands.

Contents

Acknowledgements

To tell the lost story of Britain and Ireland's wolves is an incredible privilege. My sincerest thanks go to Nigel Massen at Pelagic, who gave me this wonderful opportunity, who helped shape the book during its inception, and who was so patient throughout its writing. Thanks also to David Hawkins, who dedicated such care to the editing, and to Rachel Hudson, who designed a cover so beautiful that it surpassed everything I could have hoped for.

My deepest thanks also to all those who have been part of this journey, not only during the writing of this book but in the years preceding. My parents, who took me to the zoo as a child more times than I can count, and who took me to see my first wolves. Chris Jones, without whom I never would have discovered my passion for wolves. The Scottish Deer Centre, who allowed me to spend so much time with their wolf pack. All the family and friends, both old and new, who have been there throughout.

And the biggest thanks, once again, to Angus, who not only proofread every chapter of this book, but who supported me and it in every way imaginable, every step of the way. It would not have been possible without you.

INTRODUCTION

The Wolf of Time and (Hi)story

There's nothing like hearing a wolf howl for the first time.

It was a freezing February morning in Scotland. I was a month deep into a half-year-long stint observing a captive pack of three wolves. I knew them well by this point, and although I was aware that the residents of the town nearby were sometimes treated to a nighttime chorus, I had by now given up hope of hearing a rendition for myself.

It was an as-yet uneventful day. The wolves were clearly unbothered by the frigid temperature, their bodies protected by a thick fluff that made them look a little comically chubby, the product of thousands of years of evolution in the cold temperatures of northern America. I, on the other hand, was wearing numerous layers and thick boots, making my movements heavy and my tread clumsy. I was still cold.

The wolf in question, a seven-year-old male, had just roused from a morning snooze. He began pacing at the far side of his enclosure, as each member of the pack frequently did. Paths were worn near the fences, roads well-trodden from their patrols. For my own part I mirrored his actions, ritually pacing up and down the small woodland next to the enclosure in a fruitless attempt to regain the lost feeling in my toes, watching and waiting for something and nothing.

As the wolf paused in his back-and-forth, so did I. He looked into the distance, past the deer field with which the edges of his world shared a boundary, as if he was seeing beyond the fences and trees that obscured the view.

Just as I looked away, it happened.

For reasons unknown to me, the wolf lifted his snout into the air and let loose a long, single note.

It was over almost as quickly as it had begun. His family apparently had no urge to join in on his song.

But the effect continued long afterwards. It felt as if the air was vibrating, the chilling noise somehow reverberating through the cold mist.

I lived a cliché. This ethereal, ghostly note sent a shiver down my spine, the likes of which I had never experienced before. Like several people had walked over my grave all at once. Or perhaps a wolf.

It was not merely the thrill of finally experiencing the noise that I had long dreamt of hearing for myself. Something primal stirred in my gut, sending sparks up to the hairs on the back of my neck, as if fear of this otherworldly sound were coded into my DNA or a spell was somehow enmeshed into the song itself. Whether that tingle of uneasiness would be there had I not grown up devouring stories of young girls wearing red cloaks and little pigs in straw and stick houses, if wolves had not been reviled, feared, and vilified for centuries before that moment by my ancestors and my ancestors' ancestors, is impossible to know.

During the time I spent observing the wolf pack, I kept one eye trained firmly on another peculiar species in the park, an animal known as *Homo sapiens*. I watched and listened, observing how people reacted to seeing the Big Bad Wolf in the flesh and fur before their eyes.

Wolves have always been impossible for us to ignore. As one of the most prolific animals to walk the earth, a widespread and highly intelligent apex predator, wolves have been rich repositories of meaning for thousands of years. So long as there are people to weave narratives, tales will be told about wolves, even after the animal itself has long since disappeared.

Since the advent of the written word wolves have appeared in our laws, literature, mythologies, religions, and politics. They have permeated our languages, our idioms, and our conscious and unconscious thoughts, feelings, and ideas. Their name has come to represent anything from a person who is not what they appear (a 'wolf in sheep's clothing'), to poverty and hunger ('the wolf at the door'), an attention-seeking fib ('crying wolf'), and someone who acts alone or prefers their own company ('a lone wolf').

Margaret Atwood once wrote that 'all stories are about wolves'.[1] In other words, wolves make excellent metaphors.

But if all fiction is about wolves, so too is every story written about wolves a fiction. Their predatory nature has been adapted, twisted, and mutated in our stories and thus in our psyches. They have become beasts who kill for sport, bloodthirsty denizens of the deep dark woods. Sly and cunning beasts who stalk and devour little children, harbingers of darkness and death. This wolf of imagination is far bigger, far badder, far toothier, far more vicious, far more ferocious, far more dangerous, and far more powerful than the wolf that trod British and Irish shores so many centuries ago. But he is far more 'real' to most than the living and breathing animal upon which he is so loosely based.

Humans are hardwired to fear animals which pose a threat to us, a hangover from our prehistoric ancestors' sense of self-preservation. Our preoccupation with such creatures and their central place in our myths and legends is only natural since, as the biologist E. O. Wilson notes, 'fascination creates preparedness, and preparedness, survival'.[2] But the stories told over the centuries have distorted this intrinsic fear so that it has become vastly disproportionate to the actual danger posed by wolves – especially in comparison to other species such as bears which, despite actually being far more dangerous, have become cuddly toys and gentle honey- or marmalade-eating characters in children's stories.

Wolves die because of our myths and stories. But we do not kill wolves as wolves. We kill them as the monsters we have created.

We also fear wolves because they are a contradiction. We cannot decide whether we love and admire them, or dread and abhor them.

They are the savage monsters of our nightmares, and yet they are acutely and unsettlingly familiar. They have a fierce, humanlike intelligence with eyes that bespeak it, to the extent that the Nuxalkmc, an Indigenous people of British Columbia, 'believed that someone once tried to change all the animals into men but succeeded in making human only the eyes of the wolf'.[3] They live in socially organised packs which, bonded by loyalty, love, care, and compassion, resemble human families.[4] They are similar and familiar, considered by many to be beautiful, majestic, and even spiritual animals.

Yet they do things that we find disturbing, disgusting, and disdainful. They eat the flesh of still-alive prey. They will occasionally kill more animals than they can eat in one sitting which – from the human perspective – looks like murder for the sake of it or, worse, for the fun of inflicting pain and suffering. We sympathise far more readily with their prey than with them as predators, yet people throughout history have admired and attempted to emulate the wolf's hunting prowess, strength, and speed, even wearing their pelts to magico-spiritually 'transform' into them.

We admire wolves as beings we perceive to be truly free and wild. But these same characteristics, paradoxically, lead us to fear or hate them as entities that are entirely uncontrollable and unknowable. Throughout history, they have variously represented both gods and devils. They share 99.9% of their DNA with man's best friend, and yet they have been singled out as humankind's greatest enemy.

In their own story, the wolves of Britain and Ireland are all but silent. Mere fragments of their bodies have been unearthed. Little is known about their lives.

And so, by an ironic twist of fate, the wolf's history must be written by and through the lens of the worst-placed species to record it.[5] The most unreliable of narrators, who have caked this history and its principal character in layers of muddy misinformation and, in so doing, exonerated themselves for their massacre of its subject.

But dig a little deeper, and you will find that the history of wolves in Britain and Ireland is itself a wolf in sheep's clothing, lies told by the boy who cried wolf. A (hi)story. The tallest of tales have been repeated so frequently that they have become true in the telling and retelling. Canonical.

These myths fit so neatly with our ideas of wolves that they pass by us almost entirely unnoticed and unquestioned, and they spread ever more readily in the age of the internet. In the words of Yan Ge in her novel *Strange Beasts of China*, 'we have barely any knowledge, but are conceited enough to fill books with our ignorance anyway'.[6]

Wolves do not really howl at the moon.

The last wolf in Scotland was not a jet-black, child-eating beast killed and decapitated by a seven-foot-tall clansman in tartan.

The notion that medieval criminals could obtain a pardon with wolf tongues came from the fiery imagination of a sixteenth-century poet.

'Spitals' did not protect travellers from ravenous beasts prowling the night, but were simply a place where folk could rest their weary, frozen bones.

Wolves are not the murderers, monsters, or 'rabid droves' that they have been painted as.[7] If wolves could write a history of people, they would be far more justified in using such words about us, than we are of them.

It was not, contrary to stereotypes, their viciousness and savagery that helped wolves to become one of the most successful and prevalent species on the planet. It was their sociability and their adaptability. When circumstances change, wolves are able to change with them. Come rain or shine, snow or sun, feast or famine, they mould themselves to their circumstances. They stick together and they survive.

Wolves can live in almost any type of habitat, from tundra to desert, so long as there is enough food to eat and shelter in which to raise their pups. They can even dwell comfortably alongside humans in relative proximity to cities and suburban areas, though they have not often been treated kindly enough to do so.

Pack and territory sizes morph and shift based on the availability of space and prey. Packs generally number in the single digits, though groups of more than 30- or 40-strong have been recorded on rare occasions. They are usually made up of a breeding pair and their offspring from as many as four previous breeding seasons, who stay with the family before dispersing to find a mate and start their own packs. Relatives of the breeding male and female have also been known to be part of the pack, too, while unfamiliar wolves (usually young males) have also been observed joining packs and mating with the daughters of breeding pairs.[8]

Wolf territory sizes are hugely variable, ranging from less than 50 square kilometres to the thousands. This depends on a variety of factors, the most important of which is the availability of prey. Where prey is abundant, territory sizes are smaller since wolves do not have to travel as far to find food. In such places packs can grow particularly large, whereas in territories where food is scarce, the breeding pair's offspring may disperse to find their own territories much sooner.

These lone wolves face a perilous journey. Packs are territorial, and do not take kindly to unfamiliar wolves entering their domain. Dispersing wolves may have to travel several hundred miles to find another lone wolf to pair up with and somewhere to live. But if they can find a mate and a space to call their own, come spring they might welcome pups of their own – helpless little lumps who nestle together in their dark den, keeping each other warm, eyes closed to the world for the first few weeks of their lives. Their mother will stay with them for several months, while the male goes on forays to bring back food for his family. The next year, when the first litter is all grown up, they too will help to care for any younger siblings that come along.

The wolf's menu is diverse. Although they prefer large ungulates such as deer (red and roe in Europe), moose, and bison, wolves will also prey on rabbits, hares, boar, mice, beavers, fish, otters, and livestock animals, especially when deer are scarce or the hunting party is smaller. They will even eat berries. Being opportunistic creatures, they will also scavenge what they can find from the kills of other animals, from human waste sites (usually in places where natural prey is hard to come by), and from human burials or sites of conflict, which has not helped their reputation. When hunting ungulates, wolves will usually target the weak, sick, injured, elderly, or young members of the herd, although they are also capable of taking down healthy adults in the right conditions. Wolves are well adapted for the cycle of feast and famine thanks to stomachs capable of carrying almost 10kg of food, ensuring that they are sustained for several days while they search for their next meal.[9]

Wolves are well equipped to bring down their prey. Their well-honed senses of smell, hearing, and sight allow them to detect prey from more than a mile away. Their long legs allow them to trot comfortably at 5mph for hours at a time, but they can run as fast as 24mph for as long as 90 seconds, or even at a top speed of 35mph for short stints. Their powerful jaws enable them to keep their grip on a large, thrashing prey animal and, if necessary, support their weight should they be lifted off the ground in the struggle.[10] Their teeth are designed for tearing flesh.

Yet despite this range of weapons in their arsenal, and despite their reputation as savage and prolific killers, wolves are surprisingly unsuccessful hunters. Not only must they spend a great deal of time and energy locating an appropriate target, but they must then stalk, rush upon, and usually chase their prey, a process in which the target often escapes. Studying the hunting success of a 15-strong pack of wolves on Isle Royale, Michigan, leading wolf expert David Mech observed that of the 77 moose they targeted, the wolves killed just six.[11] The rest escaped, some outrunning the wolves (who usually give up the chase within a couple of kilometres),[12] and some standing their ground and fending the predators off. Even if a pack can get close enough to launch a successful attack, their large ungulate prey have powerful hooves and sharp antlers which can inflict fatal damage. Due to these occupational hazards, not to mention their vulnerability to human hunting and fatal inter-pack conflicts, wolf lifespans in the wild are very short. At best, a wolf that reaches adulthood can hope to live to the age of nine or ten. Most do not see so many years.[13]

Since their hunts are not often successful, wolves have occasionally been known to take advantage of unusual situations in which prey is more easily caught. Termed 'surplus killing', this phenomenon involves wolves killing multiple animals at once – far more than they can eat in one sitting. This usually involves domestic species, especially sheep, which are ill equipped to defend themselves since they have lost their wild instincts. On rare occasions wild animals will also fall victim to this practice, usually those made vulnerable by deep snow. Although surplus killing has not done the wolf's reputation any favours, it is not evidence of a savage bloodthirst. Instead, surplus killing is a logical response to an unusual situation by an animal 'programmed to kill whenever possible because it is rarely possible to kill', as experts David Mech and Rolf Peterson put it. And even if more animals than are immediately eaten are killed, they do not go to waste – wolves will return and continue to eat the surplus, or cache (bury) parts for future meals.[14]

Far from being creatures blinded by vicious rage and savagery, wolves also have complex emotional lives. Although it is perhaps anthropomorphic to describe wolf emotions in such human terms as 'happiness', 'sadness', 'anger', and 'love', those who have spent time studying these animals have frequently witnessed evidence of feelings akin to such human concepts. In his book *Decade of the Wolf: Returning the Wild to Yellowstone*, biologist Douglas W. Smith describes observing a female wolf who, seemingly against all logic and maternal instinct, left her family and her young pups after the death of her mate to undertake a week-long journey which saw her traverse inhospitable, snowy paths. For all intents and purposes, she appeared to be grieving.[15]

Wolves are also not nearly so large as we imagine, although they do have particularly big feet. Male European wolves weigh on average between 20 and 80kg and females between 15 and 55kg, with differences in latitude accounting for this wide range (wolves who live in warmer, more southerly climes tend to be smaller, while those in the colder north are generally larger). This weight is distributed across a 1- to 1.5-metre-long frame, which reaches between 50 and 70cm in height on average.[16] For comparison, Siberian huskies are around 50–60cm tall, and weigh on average between 16 and 27kg. Although *Canis lupus* is commonly called the 'grey wolf', their colouration is widely variable – from the brilliant white of wolves living in the most northerly reaches of the world to the stormy grey, shining silver, earthy brown, and auburn red of wolves further south. Their fur is double-layered – a thick, fluffy underfur provides insulation, which is kept dry by an outer layer of coarser guard hairs. This remarkable coat helps them to keep warm even in temperatures as frigid as −40°c.

Their howls, much considered by humans who may find them alternately terrifying and enthralling, are in fact a complex and highly effective communication system which can carry as far as ten miles on open ground. Howling serves multiple purposes, with perhaps the best-known function being to locate one another when pack members are separated. The purpose of howling when the pack are together and at ease is less certain – it has been interpreted by some as a way of bonding, or as a way of celebrating a reunion after time apart. Lone wolves in search of a mate and territory might also howl in the hope of finding a partner, while packs will howl at one another to warn each other of their presence and avoid meeting in the flesh for a potentially deadly confrontation. Packs will even change the pitch of their howls to make it sound as though they have more members, to ward off rival groups.

Wolves also communicate via a range of olfactory and visual signals. They scent-mark to demarcate territory, whether as a useful scent map for themselves or to signal ownership to potential interlopers. With their bodies they convey highly complex signals. They can change their tail, body, and ear posture as well as their facial expressions depending on who they are interacting with. An erect tail and ears, wide eyes and wrinkled nose, and a tense, upright posture generally signal dominance, while a lowered tail and ears and an exposed stomach indicate submission.

The taxonomy and binomial name of the wolf were first assigned in the mid-eighteenth century by Swedish zoologist Carl Linnaeus. Linnaeus placed the species under the genus *Canis* along with other dog-like carnivores – coyotes, jackals, and dogs, the latter of which he considered to be separate from the wolf – and termed the species *Canis lupus*, meaning 'dog wolf'. Even this name carries a value judgement about the animal. *Canis* ('dog') is a designation which, as anthropologist Garry Marvin points out, carries an underlying implication that '"dog" is the basic species' or the natural state, and thus that wolves and other wild canids are dogs 'gone wrong', 'wild creature[s] developing out of a domesticated animal'.[17]

Although it is now well known that wolves are the ancestors of domestic dogs, wolf taxonomy remains the subject of significant debate. Subspecies are added and discounted with remarkable frequency, and there is little consensus as to the categorisations. Many argue that attempts to be overly precise by categorising even slightly varying specimens as entirely different subspecies have led to unnecessary subspeciation,[18] an issue particularly pressing in North America where more than 30 subspecies have been identified at various times. Across Europe and Asia, current consensus suggests that there are between six and eight distinct subspecies of wolf. The majority of the European continent, however, is populated by the Common Eurasian Wolf (*Canis lupus lupus*), with only two other subspecies found in Europe: the Italian wolf (*Canis lupus italicus*) and the Iberian wolf (*Canis lupus signatus*).

Although perceived taxonomical differences may warrant the categorisation of distinct subspecies, it is accepted among biologists that 'a wolf is a wolf is a wolf'.[19] Despite minor differences in appearance, wolves across the world behave in much the same manner. So while British and Irish wolves may have had unique DNA due to thousands of years of separation from the Continent (as, indeed, do Irish hares),[20] we can safely assume that they were not so different from the wolves that roam western Europe today, both in appearance and behaviour. The extreme mobility of wolves also means that to divide wolf populations into Scottish, Welsh and English, or Irish and Northern Irish, is an arbitrary designation, convenient for our own conceptualisation of modern politically defined geographies but essentially meaningless to characterise what were likely contiguous, highly mobile populations across both islands.

This (hi)story of the British and Irish wolf begins with the very origin of the species itself, in a time of ice before the sea separated these now-islands from Europe.[21] While today we tend to think of species as 'native' or 'non-native', at this time animals moved freely across continents, treading paths long since buried beneath the waves. One such path was a land bridge connecting North America to Asia, across which it is thought that the earliest ancestors of modern-day Eurasian wolves spread westwards, evolving and adapting on the way until, a few hundred thousand years ago, they eventually became *Canis lupus*.

The wolves of early Britain and Ireland lived in a very different world, traversing variously snowy and temperate landscapes populated by bison, giant deer, bears, and even lions, hyaenas, elephants, hippos, and a now-extinct sister species of canid. While some of these species came and went as the climate warmed and cooled, Ice Age wolves were a hardy species. Their continually changing morphology and dietary preferences allowed them to adapt to the environmental, geographic, and climatic changes thrown their way. Of all the rivals with whom wolves had to contend at this time, their fiercest competitors were humans. Both Neanderthals and *Homo sapiens* arrived in this western corner of Europe during the late Ice Age, setting into motion a chain of events that would alter the fate of British-Irish wolves forever.

By 11,600 years ago the climate had stabilised, bringing with it the dawn of the Mesolithic. Britain and Ireland were now separate from the Continent, and cloaked in extensive woodland which housed a smörgåsbord of potential prey including red and roe deer, boar, aurochs, and elk, which wolves needed share with only lynx, brown bears, and humans. These humans lived increasingly sophisticated lives, not only fashioning clothing from wolf pelts but using their body-parts in rituals associated with complex spiritual belief systems. It was perhaps during this era that another key player in the wolf's fate arrived: its domesticated cousin, *Canis familiaris*. Greater still was the impact of the arrival of domestic livestock, which humans brought with them to Britain and Ireland around 6,000 years ago. With the advent of farming, humans could put down roots for the first time, staying in one place to tend to their flocks and crops. Woodland was burnt to make way for agriculture, and wolves, once respected competitors, now became a threat. The early seeds of lupicide were sown.

And yet wolves continued to serve important spiritual roles throughout the Bronze and Celtic ages. Human bodies were left out to be consumed by wolves and other predators, perhaps as sacrifices to prevent livestock depredation or as part of death rituals surrounding the deceased's spirit. Wolves also served as totem animals for the Iceni, a Celtic tribe whose coins frequently depict these animals, while statues of wolf-like creatures found at various Celtic sites across Britain may have represented lupine gods of the underworld or ancestral spirits.

When the Romans arrived in Britain in the first century CE they brought different legends about wolves, including the infamous myth of Romulus and Remus being suckled by a she-wolf. The Romans later introduced Christianity, marking a significant shift in the wolf's cultural meaning. No longer a predator to be emulated and admired, a key player in the Roman foundation story, wolves were now a spiritual enemy, a biblical symbol of vicious people, evil spirits, and the devil itself. In Ireland, however, when Christianity was introduced through contact with Roman Britain, Celtic spiritual views of the natural world were incorporated into Christian traditions. As a result, wolves were treated more favourably in Irish Christianity, frequently appearing in stories where they are not rapacious, devilish beasts, but benign animals who have been tamed by saints.

The fall of the Roman Empire in the fifth century was followed by the arrival of the Germanic tribes known as the Angles, Saxons, and Jutes, who occupied parts of Britain alongside the native Romano-British population. An age in which clothing could be woven from fibres, wolves were far less useful to people at this time than they had been in previous eras. Nonetheless, their body-parts did play a role in the spiritual lives of the pagan Germanic settlers, who used their teeth as protective amulets and their flesh and bones as part of magico-medicinal cures for illnesses. Wolves also played a positive cosmological role in the mythologies of the Germanic newcomers, who admired the wolf's ferocity and predatoriness. But missionaries from the Continent soon arrived too, intending to bring Christianity back to Britain. Within a few centuries they succeeded, and wolves were again cast in the role of devilish predators upon Christian souls.

In 1066 the fate of wolves changed once more with the coming of William the Conqueror and the Normans, who loved a spot of deer hunting. Huge swathes of the Welsh and English landscapes were swept up by the new ruling elite, who claimed these areas for their exclusive hunting pleasure. Scottish kings soon decided that they too liked the idea of hunting in their own private grounds. Only the Gaels (who maintained control of parts of Ireland) resisted the trend to parcel up the landscape, rejecting the practices of the Anglo-Norman invaders. In Britain, however, where numbers of these hunting grounds grew exponentially, wolves became a threat to royal sovereignty, preying on the same deer claimed as property by the Norman kings and lords. Thus began an extermination campaign on an unprecedented scale. Kings employed specialist wolf hunters, put bounties on wolf heads, and granted land in exchange for ridding it of its wolf population. The British and Irish wool trades were also enormously profitable at this time, providing another impetus for relentless wolf persecution or, perhaps, vice versa, with the decline of wolves allowing the wool industry to expand unhampered by their depredations. At the same time wolves were increasingly demonised in Anglo-Norman culture, ensuring their eradication in England and Wales before the end of the medieval period.

By the early modern period, the absence of wolves from England and Wales had resulted in their transformation into fascinating exotica. They became firm fixtures of zoological gardens and menageries, gawked at by visitors who enjoyed seeing these ferocious creatures behind bars, safe in the knowledge that they only roamed free far beyond the northern border and across the sea to the east. But all that changed when adventurous colonists braved the perilous Atlantic crossing to travel to the New World. Here they found the wolves of their nightmares and stories in the flesh, living and breathing threats to their livestock and, therefore, to their precarious existence. Combined with fierce religious beliefs that wolves were the devil incarnate, the threat that wolves posed to the colonial dream ensured a swift and merciless campaign of destruction that saw the American wolf population eventually dwindle to almost nothing.

Back across the Atlantic, the surviving wolves in Scotland were pushed further and further north as they desperately tried to escape from the humans that relentlessly persecuted them. But woodland was much depleted by this stage, and there were few deer to go around due to habitat degradation and overhunting by people. Eventually, tales of the 'last wolf in Scotland' began to spring up in the late seventeenth century, stories of bold deeds by daring men ridding the landscape of the lupine scourge, or brave Highland women bashing a wolf to death with a frying pan. In Ireland, meanwhile, the Gaels were once again beset by invaders from across the Irish Sea. The colonists persecuted natives and wolves alike, even calling both by the animal's moniker. It took many years, but eventually the ruthless extermination campaign succeeded. As in Scotland, the cull produced numerous tales of the 'last wolf in Ireland', stories extolling the bravery of the people – and even, in one case, the horse – responsible.

The disappearance of wolves from Britain and Ireland paved the way for the establishment of vast sheep farms, extensive estates for deer stalking, and gardens and parkland carefully cultivated by wealthy landowners to be as perfect as a painting. Wolves were thoroughly unwelcome, triggering hysteria whenever they escaped from the menageries and travelling shows that housed them for entertainment. Such hysteria was hardly surprising, since wolves were the object of vociferous mythologising by enthusiastic Victorian scholars. Even today, wolves are frequently sensationalised in the local press despite their physical absence from the landscape, their metaphorical meaning taking on new life since their extinction. Their name is now used for terrorists, criminal gangs, immigrants, and other people deemed 'undesirable' or predatory. The cultural wolf has become all-encompassing. The culmination of centuries of mythologising and demonising has transformed this shy, wary animal into the pervasive Big Bad Wolf of fairy tale and the slavering, red-eyed demon of films and novels.

But the narrative began to shift in the nineteenth century. Rudyard Kipling's (1865–1936) *The Jungle Book* (1894) presented a different side to these animals, a sympathetic portrayal that has been taken up by many storytellers since. Change was also spurred on by the first scientific studies of the wolf in the 1940s, which helped to dispel some of the more sensational stereotypes. But this new age brought with it new myths. When wolves were reintroduced to Yellowstone National Park in 1995, positive changes in the landscape were attributed to their presence and they were recast as 'saviours' of the ecosystem. This new mythology spread as readily as the old legends of Big Bad beasts, creating a new wolf of imagination which is also based on exaggeration.

It is within this context that debates about the reintroduction of wolves to Britain and Ireland are set. Some argue that wolves will 'save' the British and Irish ecosystems, controlling the overabundant deer which prevent forests from regenerating. Others suggest that that the landscape has changed too much in their absence, that wolves simply have no place in modern-day Britain and Ireland. As ever, the truth lies somewhere in between.

History is not remote. It is a narrative that is still unfolding, that we continue to write with every moment that passes. We have the chance to create new stories, tell new tales, and shape the future by learning from the past.

Throughout history, wolves deeply affected and were deeply affected by humans and human culture. Just as we have shaped their history, both as it happened and in the retelling, wolves too have shaped our lives and minds, our cultures, our languages, our psyches, our spiritual lives, and even our bodies. From the cultural ideas recorded in ancient manuscripts which continue to influence our perceptions of wolves, to the lupine place-names we have inherited in the landscape, all that has been written and thought about wolves over the past thousands of years survives in some shape or form today, whether known or unknown to us.

Uncovering these remnants of the wolf's story in Britain and Ireland reminds us that we do not exist in isolation from the past, but simply occupy the latest page in the (hi)story. Just as we cannot read the final pages of a book and have any hope of understanding what has gone before, we cannot make sense of our present nor shape our future, with or without wolves, until we understand the past.

CHAPTER 1

Wolves in a Frozen Land

The air smells cold here.

You pause, look around, seeing frozen earth dimpled by the occasional tuft of icy grass peeking through the layer of snow that blankets the ground. The land starts to rise further away, frost-covered storm-grey rock which glistens faintly in the low spring sunshine.

At first there appears to be little life. But once you are still, the steppe stops holding its breath. The bustle returns. A tundra vole pokes its head out of a hole nearby, nestled at the edge of a patch of white dryas. Catching sight of you, a great bipedal stranger, or perhaps detecting your unfamiliar scent, it abruptly turns tail and vanishes into the earth.

In the distance, from around the corner of the crag, you fancy you hear the baying of a bison. But your mind may be playing tricks on you, an auditory hallucination created by gusts of this frozen land's icy breath which echo through the rocks and blow snow over the earth, concealing the enormous hoofprints that you unknowingly passed not far from here.

Nearby is a berry bush which shows the tell-tale signs of having been ravaged by a hungry bear just waking up from its long winter snooze. Overhead you hear the unmistakable croak of a raven, followed swiftly by a cackling chorus as the flock circles and dives, painting the faint, dappled light with black streaks, like shooting stars in negative. Looking across the tundra, you see what has drawn them. The carcass of a reindeer lies in a patch of reddened ground where its blood has soaked the earth, its warmth having melted the snow. It has holes in its neck and sides from which something has been feasting, leaving behind empty cavities where organs once were.

You turn your eyes up towards the windblown crag where, even in the middle of this frozen hour of the earth's life, ice has already carved its tell-tale lines in the rock, wordlessly writing its story.

And then you see it.

Something emerges from a small hole in the rock. The cavern's entrance is well hidden – you wouldn't have spotted it had the movement not caught your eye.

The animal stops and sniffs the air, looking out across the valley. You are upwind and well hidden amongst the scree, far enough away that he shouldn't sense your presence. But still, he turns and heads back into his den. His tail swishes. Maybe you weren't so well-camouflaged after all.

You wait, eyes fading in and out of focus as you try to keep your sight trained on the hole. The light is swiftly fading. You will almost certainly lose the entrance amongst the great grey behemoth if you look away.

Several minutes pass, or it could be much longer. This place swallows time.

Then, at last, something else emerges. It is smaller, less graceful, bobbing along on little legs which it plonks down awkwardly, a comically skinny tail waving behind it. You fancy that you hear a squeaky sort of yelp as it bounds forward out of the darkness.

A wolf pup.

A Time Before Wolves

Our story begins 32 million years ago, not in Britain and Ireland but on the North American continent, with a curious genus called *Leptocyon* – the first member of the Caninae family. The eleven as-yet identified species belonging to this clade were omnivores roughly equivalent in size to the modern fox, who specialised in hunting fast and lightweight prey animals.[1] *Leptocyon* in turn gave rise to a genus named *Eucyon* (whose name means 'primitive dog'), a species comparable in size to the modern jackal, which first evolved in North America around 10 million years ago.[2]

At the dawn of the Pliocene (a period spanning from around 5.3 million to 2.6 million years ago, during which early hominins first became bipedal), the Asian and American continents became joined by a stretch of steppe grassland known as the Bering land bridge.

Members of the *Eucyon* family soon began to explore, finding more temperate conditions in these strange lands arisen from the bottom of the sea. Quickly adapting to this new environment and enjoying the tasty array of prey which they found there, free from the other canid competitors with whom they had shared the American continent, their family tree grew at an exponential rate. Multitudinous offshoots from the *Eucyon* branch blossomed and withered over the next million years, new species emerging and dying out in the cycle of life, death, evolution, and extinction.

The *Eucyon* who had remained in North America also continued to evolve until, around six million years ago, they gave rise to the *Canis* genus, a clade of dog-like carnivores. Another couple of million years later, *Canis* followed their ancestors across the Bering land bridge. Eurasia soon became, as evolutionary historians Xiaoming Wang and Richard H. Tedford put it, 'a vast playground for *Canis* evolution, setting off a wave of diversification that established the family's ultimate success'.[3] One *Canis* species that arose from this blossoming branch of the family tree was an omnivorous, medium-sized mesocarnivore named *Canis etruscus*, or the Etruscan wolf.[4] The Etruscan wolf first appeared in Europe around 2.2 million years ago and was joined by two other canid species – *Canis arnensis* and *Canis falconeri* – around 1.8 million years ago, marking a widespread expansion of the *Canis* genus so prolific that it has been termed the 'Wolf Event'.[5]

The 'Wolf Event' appears to have been prompted by another monumental dispersal known as the 'Elephant–*Equus* Event', which saw larger ruminant species (hoofed grazing herbivores) slowly oust their smaller counterparts as grassland spread and woodland shrank. Predatory species – including the canids – adapted in response, finding it more advantageous to hunt together in groups than to stalk this much larger prey alone.[6]

Each of these Eurasian members of the *Canis* family continued to evolve, ultimately giving rise to some species with which we are familiar today. *Canis arnensis* may have been part of a lineage that ultimately resulted in the modern jackal,[7] while *Canis falconeri* may be the ancestor to modern-day dholes and African painted wolves.[8] *Canis etruscus*, on the other hand, may be the ancestor to a species named *Canis mosbachensis* (or the Mosbach wolf, after the German town where remains of the animal were discovered in 1925), which in turn gave rise to the modern wolf, *Canis lupus*.[9] About as large as the petite subspecies of grey wolf found in India today, *Canis mosbachensis* first appeared around 1.5 million years ago.[10] It soon spread throughout the European continent, reaching the western annexe which would eventually become Britain by around 700,000 years ago.[11] This was the beginning of Wolf Land.

Ice Wolves

The Pleistocene, commonly known as the 'Ice Age', commenced around 2.58 million years ago and lasted until a mere 11,700 years before present. The more familiar colloquial name for this period is a little misleading, given that this epoch was not characterised by a single, continuous deep freeze. Rather, the 'Ice Age' consisted of a cycle of glacial cold snaps interspersed with warmer periods known as 'interglacials' (from which the majority of palaeoarchaeological evidence of Ice Age fauna comes), the latter of which were more or less comparable in temperature to the current climate.[12] The shifting climate presaged numerous changes in the landscape and the floral and faunal clades which inhabited it. Species moved in and out of the northwesterly outcrop of the European continent that would become Britain and Ireland as the ice sheets ebbed and flowed, sea levels rose and fell, and routes in and out were revealed and hidden.[13]

Pack-hunting canids appear to have been absent from Britain and Ireland for the majority of the Ice Age.[14] When they finally arrived during a warm spell (though still seemingly punctuated by stints of glaciation) between 800,000 and 500,000 years ago, known as the Cromerian Interglacial, they were far from being top dogs.[15] In fact, as difficult as it is to imagine today, they were one of the least remarkable predators you could encounter in Britain at this time. European jaguars, sabre-toothed cats, lions, spotted hyaenas, and bears also stalked the grasslands and woodlands, hunting mammoths, rhinoceroses, wild horses, deer, bison, and moose, alongside another member of the canid family called *Canis (Xenocyon) lycaonoides*, or the European Hunting Dog. Much larger than *Canis mosbachensis*, *Canis (Xenocyon) lycaonoides* was comparable in size to the wolves found at the northerly edges of *Canis lupus*'s range today.[16]

At this time, Britain was also home to an early species of human known as *Homo heidelbergensis*, thought to be the last common ancestor between *Homo sapiens* and the Neanderthals. The small *Canis mosbachensis* may not have been a threat to these early humans, especially in comparison to the felids and hyaenids that must have sent a shiver of fear down the spine of even the bravest of these early hominins. But the canids may have scavenged from their bodies – a human tibia found in West Sussex displays tooth marks which may have come from a Mosbach wolf.[17] Other animal bones found at this site have also been marked by both human tools and the teeth of carnivores, possibly wolves. It could be that these bones were scavenged after the humans had finished with them, or perhaps the humans butchering the animal had to fend off these pesky predators while they processed their kills.[18]

British *Canis mosbachensis* was about a third smaller than British *Canis lupus* and weighed an average of just 22kg,[19] but it may still have been sizeable enough to bring down prey equivalent in size to or even larger than itself, such as roe deer, especially if it hunted in co-operative packs like modern wolves.[20] But whether it ever dared to take on such large prey is another matter – there is debate as to whether this wolf was omnivorous or a specialised meat-eater, and whether it hunted primarily smaller or larger species.[21]

Whatever the Mosbach wolves ate, they were certainly outcompeted by other predatory species. They were no match for the sabre-toothed cats, large and powerful animals who could easily take on bison, horses, and even juvenile mammoths. Nor could they contend with the lions who were capable of bringing down bison, horses, red and fallow deer, and young rhinoceroses.[22] It was far safer for *Canis mosbachensis* to stay well away from these fearsome carnivores, lest they end up as a lion's lunch or on the business end of a sabre-toothed cat's bite.[23] Their lives were made difficult enough by the hyaenas who hunted and scavenged almost anything, probably including prey targeted or killed by Mosbach wolves.[24]

Competition between the canids was also particularly high. *Canis (Xenocyon) lycaonoides* was a highly versatile predator which was hypercarnivorous (meaning over 70% of its diet was composed of meat). It was much larger than *Canis mosbachensis*, weighing on average between 30 and 37kg, and may have hunted in packs or in pairs to bring down horses, goats, and fallow deer.[25] The presence of these large competitors had a constraining effect on the Mosbach wolf's size. There was no evolutionary advantage to growing bigger if it meant greater competition.[26]

But as time went on, the balance of power shifted. The large, hypercarnivorous felids who hunted alone – European jaguars and sabre-toothed cats[27] – started to decline. The land was no longer dominated by powerful loners hunting huge quarry, but by generalist cooperative group hunters who consumed everything from flora to fauna, from meat to bone.[28] Lions and hyaenas now ruled the roost, continuing to constrain the Mosbach wolves along with *Canis (Xenocyon) lycaonoides*, who kept their smaller cousins firmly in their place for hundreds of thousands of years until their eventual – and mysterious – disappearance from Europe around 500,000 years ago.[29]

The Rise of *Canis lupus*

Although *Canis mosbachensis* managed to survive alongside and outlast their larger cousins, they were soon to be usurped by a new competitor, another not-so-distant cousin whose invasion they would ultimately not survive: *Canis lupus*.

Canis lupus apparently evolved in Beringia, potentially from a population of *Canis mosbachensis* who had grown larger in this northerly climate,[30] allowing them to cope with both the cold and the increasing size of their ungulate prey.[31] This new species quickly spread until they had colonised much of the icy world, reaching the western edge of the European continent. They slowly but surely replaced local populations of *Canis mosbachensis* along the way, perhaps even interbreeding with them.[32] Ice Age *Canis lupus* were very similar to the grey wolves of today. Like modern wolves, they were adaptable and highly successful carnivores who lived and hunted in social groups, although they may have been somewhat more adept at crushing bone than their present-day cousins, a product of the colossal size of their megafaunal prey.[33]

The oldest *Canis lupus* remains discovered in Britain belonged to a wolf who lived in and around Pontnewydd Cave in Clwyd around 225,000 years ago, during a warm spell known as the Aveley Interglacial. This cave was also home to Neanderthals at some point during the same period (though probably not at the same time as the wolves), who left behind many of their stone tools and even some of their teeth – the oldest hominin remains yet discovered in Wales. The presence of bear, leopard, horse, roe and red deer, mountain hare, beaver, and wood mouse remains, along with those of the wolf and

Neanderthals, indicate that the landscape around Pontnewydd Cave at this time was composed of primarily open woodland.[34]

An increase in temperature towards the end of this interglacial created a landscape dominated by open grassland. Elephants, bison, mammoths, wild horses, aurochs (the enormous ancestors of domestic cows), rhinoceroses, and red, roe, and giant deer (7-foot-tall colossi with a 12-foot antler span) grazed here, preyed on by lions, bears, hyaenas, and leopards who launched their ambushes from within the woods that interspersed the grassland.[35] With the disappearance of most of the solitary big cats, *Canis lupus* was a more prominent member of the carnivore guild than its smaller ancestors.[36] Yet despite this, the wolves of this period decreased in size as the interglacial went on, perhaps due to the fluctuating climate, the isolation of Britain from the Continent due to rising sea levels, or competition from the remaining larger predators.

At this time lions were common and particularly big, capable of taking down anything from deer to juvenile mammoths, elephants, and woolly rhinoceroses.[37] In response to competition from these impressive felids, wolves appear to have steered away from targeting large prey species, their teeth suggesting that they adapted to tackle smaller prey instead.[38] Though targeting smaller quarry may have brought wolves into conflict with the (albeit rarer) leopards, and hyaenas likely continued to hunt and scavenge anything they could get their paws on (including prey brought down by other carnivores), it appears that there was plenty of meat to go around. Lack of breakage on the teeth of wolves from this period indicates that they were not forced to consume their kills quickly to avoid theft, while a lack of wear implies that they were not forced to eat the whole carcass, bones and all, before finding their next meal.[39] They also appear to have taken advantage of the vegetarian foodstuffs which the warm grassland environment provided, another useful strategy for avoiding conflict with other predators.[40]

But of all the predators with whom *Canis lupus* was forced to compete, none was more deadly than the Neanderthals. Found commonly throughout much of Britain, having appeared sometime during the colder period which preceded the Aveley Interglacial, Neanderthals were extremely successful and diversified predators capable of hunting and killing prey of all sizes.[41] They were powerful members of the carnivore guild and were similar to wolves in a number of fundamental ways, bringing the two into direct competition. Contrary to stereotypes, Neanderthals were intelligent and adaptable and, like wolves, they lived and hunted in groups. Wolves and Neanderthals both also scavenged opportunistically and were able to diversify their diets if the environment allowed or necessitated it, although there is evidence that the Neanderthal diet contained a lot of meat. Both species used caves for shelter, and it seems likely that Neanderthals traversed significant distances, at least in colder climates, as do wolves.[42]

Neanderthals themselves apparently preyed on and exploited wolves, as is indicated by wolf bones bearing chop marks which were found in a cave on Jersey.[43] Yet their relationship with wolves and other predators may have been based on more than simple exploitation or competition. Research has indicated that Neanderthals maintained close and complex relationships with other predators – they were seemingly 'deeply attuned' to these animals, with predators playing an integral role in Neanderthal culture and society.[44]

A cold period that began around 190,000 years ago forced Neanderthals to move southwards and out of Britain, whose landscape had been transformed into a cold, steppe grassland. Little is known about the animals which inhabited Britain at this time, though it appears the faunal clade was very limited, comprising wolves, horses, red and arctic foxes, brown bears, and other small mammals.[45] These wolves lived a harsh existence, their highly worn teeth suggesting prey scarcity which forced them to consume all parts of the carcass lest they went hungry.[46]

Around 130,000 years ago, temperatures began very rapidly to increase. It stayed particularly warm for 15,000 years, a period known as the Ipswichian Interglacial. Sea levels rose along with the temperature, submerging the land connecting Britain to Europe, leaving it isolated. Some mammals who had vacated the area when temperatures were colder were quick to recolonise the island before it was cut off, while other, larger animals may have made the journey via the water. Too slow off the mark and with no means of crossing once the waters had risen, on the other hand, Neanderthals were not to return until many tens of thousands of years later.[47]

But the wolves found in Britain during this period still had to compete with lions, bears, and a particularly abundant population of hyaenas which, as in previous times, constrained the wolves' size.[48] Nonetheless, this mosaic landscape was occupied by red, roe, fallow and giant deer, bison, wild boars, rhinoceroses, elephants, and even hippopotamuses,[49] an abundant prey guild which kept the wolves and other predators well fed.[50]

But the good times were not to last. A period of extensive glaciation which began around 110,000 years ago saw much colder temperatures set in. Very few records of the animals living in Britain at this time survive today. We know nothing of the wolves that may or may not have continued to live here. Maybe they stuck out the deep freeze, or perhaps they travelled south in search of better fortune. They do reappear in the slightly warmer period which followed around 100,000 years ago, in a mosaic landscape of woodland and open grassland roamed by hyaenas, roe and red deer, bison, woolly mammoths, rhinoceroses, and elephants.[51] They then disappear once more during another cold snap around 90,000 years ago, only to re-emerge again around 80,000 years ago. This was a slightly more temperate phase, though it was still cold by today's standards, with highs of just 11°C and lows of −30°C. At this time, Britain was dominated by open tundra traversed by steppe bison and reindeer.[52]

Wolves came into their own during this period. They were finally free of competition not only from Neanderthals but lions and hyaenas too, who appear not to have made it back over before Britain (apparently) became an island once more.[53] The only downside was that brown bears had managed to return, and they were so big that their remains have been mistakenly identified as belonging to polar bears.[54] Otherwise, wolves needed to compete only with foxes and wolverines, who were more of a nuisance than a threat. Nothing a swift snap of the jaws couldn't solve.[55]

Living in cold conditions and unconstrained by the larger predators of earlier eras, these wolves adapted by increasing in size. Reaching almost the average weight of *Canis lupus* today (41kg), at this size an adult steppe bison could have been a pack's supper, even despite weighing an enormous 672kg on average.[56] The irony, however, is that while they were free from the competitors who for so long had limited their options, the wolves had little choice but to prey on these previously off-limits giants. Steppe bison were an especially valuable resource since they were a seasonal delicacy, solely available when herds migrated into Britain. The only other large ungulate found in Britain at this time was reindeer, another migratory species.[57] Besides these two animals, it was slim pickings. The hares and voles found on the tundra would barely make a dent in a hungry wolf's empty stomach, let alone feed a whole pack.

Despite the disappearance of lions and hyaenas, the harsh environment will have brought the wolves into conflict with species with which they had previously had little quarrel. Brown bears will have been forced to adapt to the dearth of plants by tending towards carnivorousness,[58] and they were certainly big enough to drive a wolf pack away from their hard-earned meal.[59] Wolves were likewise forced to adapt to the impoverished environment by consuming anything they could find, though scavenging was mercifully much less risky in the absence of hyaenas.[60]

These harsh conditions and tough competition for food took their toll. The teeth of the wolves who lived during this period were better adapted to slice flesh, suggesting that they were consuming prey quickly to avoid theft by other carnivores, including rival groups of their own species and even members of their own packs. Their teeth were also adapted for cracking bones which, along with high levels of breakage and much wear and tear, suggests that food was so scarce that these wolves were forced to expend energy on extracting marrow, eating the less delectable bone, and even consuming frozen carcasses.[61] Reindeer antlers were apparently the favourite food of at least one pack of wolves who denned in a Somerset cavern named Picken's Hole, where over a hundred pieces of antler dating to this period have been found. Both male and female reindeer shed their antlers yearly but females and juveniles drop theirs in the spring, precisely the same time of year that growing wolf pups are weaned and in need of nutritious chew-toys. Dents and tooth marks on the antlers found at Picken's Hole suggest that both adults and pups alike enjoyed this protein-rich snack.[62]

Another period of extensive glaciation between around 75,000 and 57,000 years ago leads to the disappearance of wolves from the archaeological record once more. A 30,000-year stint of slightly milder but highly changeable climate then followed, during which the landscape was dominated by open grassland populated by herbaceous flora, berry plants, scrub, and birch trees. These plants sustained a wide range of herbivores including woolly mammoth, woolly rhinoceros, steppe bison, wild horse, reindeer, red deer, giant deer, mountain hare, arctic lemming, tundra vole, and a curious little species of ground squirrel called the red-cheeked suslik. The smaller carnivores who preyed on these animals included stoats, polecats, red foxes, and arctic foxes, while the clade of larger predators comprised wolves, brown bears, and potentially sabre-toothed cats.[63] Lions (possibly both cave lions and the species of lion we know today, *Panthera leo*), hyaenas, and Neanderthals also returned thanks to the reappearance of a land bridge connecting the European mainland to Britain.[64]

Wolves and Early Humans

Though their march back into Britain saw Neanderthals resume their place at the top of the hierarchy, they were soon to be displaced by an even more formidable predator. At some point between 44,000 and 41,000 years ago *Homo sapiens* appear to have arrived in Britain.[65] Just a few thousand years later, the Neanderthals disappeared. Not only did these newly arrived humans compete with (and outcompete) Neanderthals, but their preferred prey also brought them into competition with the wolves, hyaenas, and lions found in their new home,[66] who were all drawn to Britain for its abundant prey populations.[67] These humans were prolific hunters, with intimate knowledge of the animals they hunted. Unlike Neanderthals, whose hunting strategy was to accost and kill single animals, *Homo sapiens* could kill entire herds by ambushing them as they crossed rivers or passed through narrow ravines.[68]

Wolves appear to have been particularly numerous in southern England at this time, judging from the plethora of remains found in a group of caverns nestled in a limestone hill in Plymouth, known as the Oreston Caves. Excavated by quarrymen in the first half of the 1800s, fifteen baskets-full of hyaena, wolf, fox, horse, deer, and aurochs remains were uncovered within just one of these caves. Another cave contained two cartloads of bones, but these were disposed of by workers before the man who was studying this cave, Joseph Cottle, had even arrived at the scene. Despite these extensive disposals, Cottle subsequently collected 40 wolf jaw bones alone. Being a religious man, he determined that they belonged to animals that had perished in the Flood.[69]

Knowledge has since moved on, and we now have a much more accurate picture of the wolves of that time. Once again, they were outcompeted by lions, hyaenas, and both species of human. In combination with the warmer weather and the more diverse flora and prey fauna that went along with it, the reappearance of the wolves' old rivals led their bodies to shrink in size once more and prompted them to turn once again to vegetal foodstuffs, and even insects, for some of their meals. When meat was on the menu, it came only from smaller prey species. There was no succulent woolly rhinoceros or mammoth for these wolves. These species were best left for the hyaenas, who became progressively more difficult to compete with as time went on, not only increasing in numbers but also growing larger and stockier. Nor was it wise to take a chance on a reindeer or a bear cub, likely the prey of choice for the lions. Perhaps even worse were the sabre-toothed cats. They were fast runners who may have hunted in groups, bringing them into direct competition with the pack-hunting wolves.[70]

A wolf jaw dated to this period caused quite a stir when it was first discovered by nineteenth-century geologist Hugh Falconer, who unearthed this mandible in a cavern on the Gower known as Spritsail Tor. Described by Falconer as a 'hyaenoid wolf',[71] it was subsequently attributed to the species *Lycaon* by naturalist Richard Lydekker due to the presence of 'very remarkable tooth' with a distinct anomalous projection known as a talon cusp. Lydekker named this species *Lycaon anglicus*, emphatically noting that 'there is no question but that the specimen is specifically distinct both from *Canis lupus* and *Lycaon pictus*' (the painted wolf).[72] The tantalising possibility of a British species of *Lycaon* was subsequently dashed in the early twentieth century, however, when geologist Sidney Reynolds determined that the jaw belonged simply to a 'somewhat abnormal wolf'.[73]

Around about 29,000 years ago the world entered a lengthy deep freeze which, aside from a few thousand years of slight alleviation, was largely characterised by extreme cold until around 11,700 years ago. Even in summer, temperatures never reached above the single digits.[74] An ice sheet covered much of Britain at this time. At its greatest extent, it enveloped all of Ireland and Scotland, most of Wales, and a large portion of northern England, leaving only the very tops of the mountains exposed. The rest of the landscape was bare and cold, cut through by lakes and streams from the glacier's meltwater and chilled by bitter winds. Only animals well adapted to the cold could survive these extreme conditions, and even then they occupied only the southernmost areas of the tundra, perhaps only on a seasonal basis.[75] With the lions and hyaenas dwindling in number and eventually disappearing with the coming of the cold,[76] wolves were left to compete with only brown bears and foxes. This predator guild dined on hares, woolly mammoths and rhinoceroses, reindeer, red and roe deer, wild horses and, most tellingly of the severe climate at the time, musk oxen, a species adapted to extremely cold conditions.[77]

Though previously thought to have abandoned Britain during this time, *Homo sapiens* may in fact have made forays to these northerly parts of Europe,[78] making use of the local animals for both their meat and their pelts. Wolf fur in particular is an excellent insulator, given that it has a dense layer of downy fur which is overlain by wiry guard hairs that protect the soft inner fur and trap heat. This keeps wolves warm even in sub-zero temperatures, a fact well known to Inuit peoples who still utilise their fur today, particularly on hoods to protect the wearer's face from frostbite.[79]

There exists a fine-grain measurement for how insulative an item of clothing is, called the 'CLO value'. The baseline for a naked person is 0, while a modern-day winter parka has a CLO value of 0.7. Husky fur has a value of 4.1 or above, while wolf fur can reach up to 7.5. In late Ice Age Europe, early humans are thought to have been exposed to temperatures as low as -20°C and an even colder wind chill, which can cause frostbite in just half an hour if skin is exposed. Early hominins did not know about CLO values, but they do seem to have been aware that wolf fur could protect them from the cold. Lupine remains are often found at European sites used by Neanderthals, and even more frequently at those used by *Homo sapiens*.[80] A pelt at this stage of the late Ice Age would be very large and therefore very lucrative, the wolves possibly weighing almost as much as 40kg.[81] Whether they ate the meat of the wolves that they hunted for this purpose is unknown.

Some of the earliest records of *Canis lupus* ever found in Scotland date to this severe cold phase. Scottish wolf remains in general are exceedingly rare, but there are still tantalising traces of the animals that used to live here. In a network of caverns that punctuate an imposing Sutherland mountainside called Creag Nan Uamh (the 'Crag of the Caves'), the presence of pup remains suggest that wolves denned in the caverns for which the mountain is named.[82] But it was not just the spacious caves that may have attracted them. Hundreds of pieces of reindeer antler were washed into these caves between 48,000 and 22,000 years ago, the majority belonging to female reindeer and the rest to juvenile males. These antlers, along with a path carved into the rock by the clatter of hooves ascending to the top of the hill, suggest that there was a calving ground further up the slopes, a well-stocked living larder upon which wolves could prey.[83]

A hollow in a nearby hillside is named Coire a' Mhadaidh ('corrie of the wolf'), suggesting that wolves continued to frequent this part of the Scottish landscape for millennia.[84] Wolves doubtless traversed many of Scotland's hills and glens during the thousands of years that they called this landscape home, their howls drifting over misty lochs and mountain passes, yet they have left behind barely a whisper to indicate that they were ever here at all.

Ireland's wolves have likewise left only fragmentary echoes of their existence, since the often-acidic soil here is not conducive to the long-term survival of organic matter. The oldest wolf remains found on this island date to the same period of extreme cold as the earliest Scottish remains, comprising two jawbones found in caves in Waterford and Cork. The population of wolves to which these individuals belonged had likely recolonised Ireland during a brief stint when the ice sheet retreated, and were joined by other species previously forced out of the island during periods of extreme glaciation, such as arctic foxes, woolly mammoths, and reindeer.[85] Wolves probably survived by predating on the seasonally migrating reindeer herds, as well as hare and other small mammals such as lemming and stoat.[86] We owe much of what we know of Ireland's Ice Age faunal clade to these wolves. They apparently used caves (such as those in which their jawbones were found) for breeding and shelter during stretches of severely cold weather, dragging the corpses of their prey into the stony depths to be consumed at their leisure.[87]

Temperatures first began to rise significantly around 14,000 to 13,000 years ago, during a period known as the Windermere Interstadial (in Britain) or the Woodgrange Interstadial (in Ireland), which lasted for a few thousand years. Animals typical of more temperate (though still cold) climates were found in Britain and Ireland at this time, where summer temperatures reached a balmy 16°C or so.[88] Treading Ireland's grasslands were giant deer, reindeer, and perhaps red deer, preyed on by brown bear and wolf.[89] The mixed steppe and woodland environment of Britain housed *Homo sapiens*,[90] red foxes, arctic foxes, brown bears, lynx, and wolves, who had once more assumed their place in the upper echelons of the predatorial clade following the disappearance of hyaenas and lions. These predators preyed on mountain hares, lemmings, voles, pika, beavers, mammoths, wild horses, reindeer, red deer, aurochs, and saiga antelope.[91] Saiga antelope are smallish ungulates with rather unfortunate-looking noses. Today, they are found on the Eurasian steppe in dry and subdesert conditions, though they historically inhabited forested steppes in more northerly parts of the world. During this warm and dry interstadial they spread far and wide, and could be found all the way from Alaska to Britain.[92] Their arrival in Britain provided a succulent new source of meat for the wolves' menu.

A wolf pack living in a Cheddar Gorge cave known as Sun Hole brought saiga antelope carcasses back to their den to feast on, along with hare, beaver, reindeer and wild horse.[93] Humans also occupied this particular cave at one time, although the *Homo sapiens* remains also found there probably did not belong to these residents, but were scavenged by wolves from another cave on the opposite side of the gorge.[94]

This second cavern, called Gough's Cave, is one of the most important British sites of the late Ice Age. The famous Cheddar Man, a skeleton of a male *Homo sapiens* who died around 10,000 years ago, was buried here. But the Cheddar Man was not the only human resident of this cave. Four thousand years earlier it was inhabited by a group of people who appear to have used it for several generations as a camp, judging from the large number of stone tools they left behind – the most, in fact, of any British cave site occupied by Stone Age people.[95] A range of faunal remains also found in the cave appear to have been the byproducts their meals, leftovers that may have been scavenged by wolves when the people left for pastures new at various intervals throughout the year.[96]

The sheer number of artefacts and bones uncovered in Gough's Cave provide valuable insight into both the animals that lived during this time, as well as the people who hunted them. It would have yielded even more, but when the site was first excavated in the early twentieth century, numerous tea-chests full of bones were simply thrown away because they were 'duplicates'.[97] Tea-chests are not small – neither, therefore, was the number of bones that were lost.

The people who dwelled in Cheddar Gorge all those years ago belonged to the western European Magdalenian culture, which lasted from around 17,000 to 11,000 years before present. As climatic conditions gradually improved, the Magdalenians travelled westwards and eventually crossed to Britain via Doggerland, a large swathe of land which emerged from the bottom of the North Sea to connect Britain with the European continent.[98] The Cheddar Gorge sites provide the earliest known evidence of Magdalenian occupation in Britain.

The Magdalenians were semi-nomadic hunter-gatherers who lived in natural shelters such as caves in the winter, but utilised tents during the summer. Their ability to remain settled for parts of the year suggests that there was an abundance of food available, which precluded the need to travel far to find their next meal. They made use of weapons and traps when hunting their prey, which constituted primarily herd animals such as wild horses and red deer. They were not only capable of carving bone and ivory to construct tools, weapons, and jewellery, but they could also engrave geometric patterns and artistic representations of animals into these artefacts and into stone.

As with their Neanderthal predecessors, the Magdalenian peoples' relationship with wolves – and nature more widely – likely went beyond mere utility and survival. But the manner of this relationship is murky. Predators are exceptionally rare in Palaeolithic art despite their prevalence in the landscape, none more so than the wolf. Of all known cave paintings that have yet been discovered throughout Europe, fewer than ten definite depictions of wolves have been identified, with a handful more of spurious – but potentially still lupine – identity.[99] Most surviving cave art is found in mainland Europe, particularly in France, with an engraving of a large stag carved into the wall of Creswell Crags near Worksop one of the few surviving examples in Britain.

Why these Palaeolithic people focused almost exclusively on prey rather than predatory animals in their art is a mystery. It has been attributed by some to a concern with food, perhaps serving a magical function intended to improve hunting success or ensure that herds were abundant and healthy. The paintings could even have been used to teach children about the animals they would grow up to hunt, imparting knowledge of their behaviours and how to successfully kill them.[100] Others, following eminent anthropologist Claude Lévi-Strauss, suggest that early humans were primarily concerned with certain species not because they were 'good for eating', but because they were 'good for thinking'.[101] But in the case of the wolf, which throughout history has proved one of the most fruitful species to 'think with' and about, this theory leads to more questions than answers. It is hard to imagine that the wolf was not central to Magdalenian mythology and storytelling,[102] just as it is today. Wolves were a top predator which, like early humans, lived in tight-knit social groups. Both species targeted the same prey, and wolves potentially posed a threat to human life. Wolf-tooth pendants found in Palaeolithic burials in Europe attest to the perhaps spiritual relationship of humans with wolves.[103] It seems, therefore, that for some reason wolves appear to have been '"good to think [about], but not good to draw"'.[104] Perhaps to draw them was even taboo.[105]

Archaeologist Matthew Beresford argues that those early humans navigating the perilous natural environments of prehistory – lands of 'physically powerful and threatening animals' with 'infinitely more ferocity and much more strength' than *Homo sapiens* – would 'try to imitate wild animals in order to survive and find food'. This perhaps led them to 'attempt[] to look and feel like the wolf by wearing its pelt or its teeth', or to 'involve [...] the wolf in ritualistic ceremonies'.[106] Beresford contends that one such ritual is depicted on a woolly rhinoceros bone found in a Worksop cavern known as Pin Hole Cave, upon which has been carved an animal-headed human or a human wearing an animal mask, who seems to be engaged in a ritual dance.[107] But the mask or animal head is unfortunately nondescript, its identity elusive.

While some remains of the animals that fell prey to the Magdalenian people, like this woolly rhinoceros bone, were fashioned into more remarkable objects, others received less decorous treatment. Many of the wild horse and red deer remains which dominate the Gough's Cave assemblage are riddled with grooves, stories written upon their bones which tell of skinning, the disarticulation of limbs, the removal of flesh from bones, the extraction of sinews and tendons to begin new life as thread or rope,[108] and the smashing of skulls and bones to reveal a delicacy of brain tissue and bone marrow.[109]

Humans were not the only species that left remnants of their meals behind in Gough's Cave. Other animal bones found towards the back of this cavern display the tell-tale signs of having been gnawed by a canid, perhaps a red or arctic fox, or a wolf. Remains of all three species have been found in the cave,[110] and it is possible that these canids moved into the cavern during periods when the humans had vacated it, scavenging from what the people had left behind and taking their spoils to the back of the cave where they could enjoy their feast undisturbed.[111] Or perhaps the masticator responsible did not have to wait until the cave was unoccupied to enjoy their scraps. The canid remains found in this cave may not have belonged to a wolf or a fox at all, but to one of the earliest known domestic dogs ever found in Britain. This dog may have been a hunting companion, helping humans to bring down the wild horses which were butchered in Gough's Cave.[112]

As the climate warmed the Magdalenian people pushed farther northward, reaching as far as Langcliffe Scar in the Yorkshire Dales. This area of Yorkshire was particularly attractive to both human and lupine species, since herds of large herbivores – reindeer, aurochs, and wild horse – were likely using a nearby spot as a calving ground during the spring and early summer, providing a glut of vulnerable prey. Humans hunted the calves for their skin (a particularly useful source of very soft leather), while wolves will have hunted the young animals to feed their own growing pups, some of whom appear to have been denning in a craggy cavern set in a dramatic cliff of Langcliffe Scar known as Victoria Cave.[113] Although some of the gnawed and cracked bones of the herbivores left behind in this cave are undoubtedly the products of wolf kills, or body-parts scavenged from carrion which they had fortuitously stumbled upon, wolves appear to have swiped bits and pieces from human kills with relative frequency. Both aurochs and wild horse bones encountered in this cave bear the characteristic dents associated with the teeth of a wolf, yet they are also inscribed with the tell-tale grooves that only a human's tools could leave behind.[114]

A very similar story is told by the remains found in another group of caves in North Yorkshire, which are set in a cliff with the delightfully bizarre name of Giggleswick Scar. One of these caverns, known as Sewell's Cave, yielded horse and reindeer bones which had been scratched by humans and gnawed by wolves, who retreated to this den with their stolen booty.[115] Just like their British cousins, wolves in Ireland also appear to have been consummate thieves. A cavern in Waterford called Kilgreany Cave contained a giant deer tibia which had apparently been both gnawed on by a wolf and split open by a person in search of the marrow, though not necessarily in that order.[116] That wolves actively scavenged from their kills suggests that the humans of this period were significant predators in the ecosystem. In fact, it might be that wolves were not pilfering from human kills simply because it was an easy meal, but because they were driven to do so by competition for prey from these expert killers.[117]

Following this warm phase, around 12,900 years ago there was a (geologically) brief cold snap which lasted for a mere 1,200 years. Known as the Younger Dryas, or the Loch Lomond Stadial in Britain and as the Nahanagan Stadial in Ireland, this era saw the slow but sure melting of the ice sheet that had enveloped Britain and Ireland interrupted by a period of renewed advance, which forced warm-weather animals out of the homes that they had only recently moved into. Wolves may have been the only large carnivore present in Ireland at this time, if they did not retreat from these cold, northerly climes. The environment was perhaps too harsh to allow a viable population of brown bears to survive in this isolated area, with only reindeer, a declining population of giant deer, and small mammals such as hare to prey on.[118]

But around 11,700 years ago, temperatures began to climb once more. Summertime saw highs much closer to those we experience today, at 20°C. Open woodland began to replace the tundra landscape.[119] Animals travelled further and further north as the ice sheet began to shrink once again, crossing from mainland Europe to Britain via Doggerland. Though it was slowly sinking beneath the waves as melting ice turned to sea, Doggerland would not be fully submerged for several thousand more years.[120]

It is not clear whether sea levels were ever low enough to create a path between Ireland and Britain at this time, although the absence of animals such as frogs, snakes, and voles in Ireland does appear to suggest that there was no bridge for them to cross over. Instead, the large animals that recolonised Ireland after the end of the Ice Age (if they had not managed to cling on during the deep freeze) – brown bears, red deer, wild boar, and wolves – may instead have swum across from the mainland,[121] the latter of whom can travel distances of up to 8 miles by water.[122] The Irish Sea was narrower then than it is today, and was quite possibly shallower and interspersed with hummocks of dry land due to lower sea levels, facilitating an easier crossing.[123] It is even possible that some animals made their way to Ireland from northern France, making use of islands which may have punctuated the 500km expanse of the Celtic Sea. This raises the intriguing possibility that the Irish and British wolves wiped out just a few centuries ago were not so closely related as one might expect, just as the DNA of modern Irish pine martens is far more closely matched to their Spanish cousins than their British neighbours.[124] As with the Irish hare, whose genetic make-up is distinct from hares in Britain and on mainland Europe,[125] the DNA of Ireland's lost wolves may also have been unique.

With the end of the Younger Dryas and the warming of temperatures once more, the Ice Age came to its conclusion. A new epoch was dawning, one that would see humans shape the fate of both wolves and the entire natural world on increasingly unprecedented scales. The Holocene was beginning.

CHAPTER 2

Wolves at the Dawn of History

A small group of people is gathered around a fire at the edge of the lake. Children have drifted off to sleep on scattered furs while the adults talk and share songs, their faces animated by the flickering flames and the fire's reflection in the water. A furred animal lies nearby. At first, in the dim light, you mistake it for a small wolf. But its ease in human company as it lazily chews on a bone tossed its way bespeaks its domestication.

Your eyes settle on a grizzled man who sits apart, head bent to his work. A simple knife in one hand, he is carefully separating skin and fur from flesh and bone. The pelt is grey, mottled with fox-red undertones which catch the light. It used to belong to a wolf.

The man concentrates on his task, at pains to preserve both the fur which will keep his children warm this winter, as well as the body of the animal who gave its life to help them survive. His face is hard to read as he quietly completes his work.

Now parted from its flesh, the wolf's pelt is carefully laid out to dry, soon to become a piece of clothing or bedding. The man returns to the disrobed animal's body, picking it up as gently as if it were his own child. He lowers his head as he bends to carefully emplace the animal in the lake, still fiery with the echo of the nearby flames. He thanks the wolf for its sacrifice in a ritual passed down to him by his parents, who learned it from their parents before them. He knows what he has taken, the price that has been paid in the exchange of life for life.

Wolves in a Warming World

The first five-and-a-half thousand years of the Holocene, known as the Mesolithic (from around 11,600 to 6,000 years ago), saw humans learn to bend natural substances to their will. They mastered the use of flint to make tools and weapons, a development that cemented their place atop the food chain despite being far fewer in number than other top predators in Britain. They subsisted on a variety of food from both floral and faunal sources, which also provided valuable materials for crafting clothing and jewellery, structures to live in, weapons to hunt with, and items serving aesthetic and spiritual purposes. On forays to the coast they hunted seals and birds, plucked molluscs from estuaries, and trapped fish on their migrations. While inland they tracked deer, boar, aurochs, and any other animal whose body they could put to use.[1]

As temperatures increased at the dawn of the Mesolithic, the tundra of Britain was initially replaced by scrub. The vast spread of woodland soon followed, which by around 8,000 years ago had come to dominate the landscape (although its extent remains a matter of some debate). These changes were favourable to species which enjoyed warmer temperatures and thrived in woodland habitats, including roe and red deer, wild boar, elk, aurochs, beaver, hedgehog, and various species of vole, which in turn were preyed on by wildcat, pine marten, lynx, bear, fox, and wolf.[2] Reindeer, wild horse, and giant deer also returned, but although they managed to eke out an existence for a few thousand years, the climatic and environmental changes proved too much for them, and they disappeared a few thousand years later.[3] Even so, the 6,500 to 8,000 wolves estimated to have lived throughout mainland Britain during the Mesolithic still had plenty on their menu. Red and roe deer were particularly abundant, though the wolves did have to share them with a modest number of humans, as well as abundant lynx and a prolific population of brown bears.[4]

Ireland's landscapes also changed following the end of the Ice Age. The tundra and scrub were slowly but surely replaced by open woodland which became denser as the Mesolithic went on. The seeds of trees and scrubs were carried over the Irish Sea by birds, tides, and wind, although around one third of the tree species found in north-west Europe at this time did not make their way over the watery barrier to Ireland.[5] As in Britain, the giant deer and reindeer that had recolonised Ireland towards the end of the Ice Age disappeared soon after the Mesolithic began, while roe deer and aurochs never made it over the Irish Sea. Red deer, which were possibly killed off in the cold snap before the Ice Age's end, may not have recolonised until as recently as 5,000 years ago.[6] Hence, Ireland's wolves had far fewer prey options than their cousins across the sea. Rather than the large ungulates that their neighbours enjoyed, they survived on hares; migratory fish such as salmon; and birds like bitterns, cranes, and auk; along with vegetarian foodstuffs such as berries.[7] The arrival of boar around 9,000 years ago (possibly the doing of humans, who began to settle Ireland about a millennium earlier) supplemented their diet considerably. These wild pigs may have become so prolific that they alone could have supported a population just shy of 2,000 wolves,[8] although the wolves did have to share them with bears and lynx.[9]

Despite the fact that wolves could have been highly prolific at this time, there is very little evidence of the thriving wolf population that Britain may have sustained, and nothing definitive to prove their presence in Ireland at all. Caves, where faunal remains are often found, had become prime real estate after the glaciers which expanded towards the end of the Ice Age damaged their entrances. Of the few caves that remained open, a wolf would be lucky to find one that remained dry in the wet Holocene climate. Even if they did locate a cosy cave to crawl into, they might have to fight off a group of soggy, irritable humans to claim it.[10]

Mesolithic-age archaeological sites are also extremely rare in general, especially in the acidic soil of Ireland and Scotland. Furthermore, since most archaeological digs are undertaken at sites where human settlements used to be, evidence is skewed in favour of the prey species that were most frequently hunted by people (those which were eaten),[11] and inevitably excludes any animals whose bodies were processed and/or discarded elsewhere.[12] Compounding these difficulties is the fact that it is often impossible to identify wolf remains with certainty given their morphological similarity to domestic dogs, which had arrived in Britain and Ireland towards the end of the Ice Age.

One such mysterious canid was unearthed in Derry at a site known as Mount Sandel. This 7,000-square-metre camp overlooking the River Bann was inhabited around 9,000 years ago. Mount Sandel is famous for its ten circular structures which are the first known 'houses' built in Ireland, while an 'industrial area' nearby was dedicated to flint-working to make weapons and tools. The inhabitants of this site, possibly a single family, enjoyed hazelnuts, waterlilies, and wild pears along with a range of meat and fish which they cooked in a multitude of hearth pits clustered around their houses.[13] Over 2,000 fragments of burnt animal bones have been found in these pits, most belonging to fish but some to boar, hare, wildcat, eel, and various species of bird, as well as the remains of a wolf or dog which may have come from an animal whose pelt was removed by the site's inhabitants.[14] Given the rarity of domestic dogs in Europe at this time, historical geographer Kieran Hickey has even suggested that this animal was a wolf who had been reared from a pup by the people who lived here.[15]

People in Ireland had already altered the environment far more than other species were capable of. Unlike in Britain, with its greater variety of prey species whose populations were replenished by the land bridge still adjoining Britain to the Continent, humans living in Ireland had far fewer options.[16] In the absence of a crossing to facilitate the movement of prey, these humans decided to give certain species a lift across the sea. This had profound and complex effects on the ecosystem and the faunal community structure. Boars in particular, masters of ground disturbance, may have increased diversity in Ireland's woodlands, rendering once-impenetrable forest (which was less heavily grazed upon than that in Britain[17]) far more open. But unfortunately for the people who had taken so much trouble to transport them over the waves, these wild pigs will have been a firm fixture on the wolf's menu, which in turn allowed numbers of the predators to increase. Less favourable to the wolves was the introduction of other predators. Both wildcat and lynx may have been introduced by people for their fur, the latter of which, as adept predators of wild boar, will have directly competed with the wolves for this food source. Even bears may have been introduced to Ireland by humans, perhaps for spiritual reasons. In turn, the small mammals and birds that wolves had previously preyed on were now hunted by a more diverse and numerous suite of predators.[18]

Although they had access to a broader range of prey than their neighbours across the water, people in Britain were also manipulating their environment in new ways. The relationship between humans and their prey had changed since the Ice Age – following the loss of large, slow-moving megafauna which could be brought down with spears, people turned to hunting smaller, more elusive prey with bows and arrows.[19] To make such prey easier to target, it is thought that Mesolithic people in Britain (and, to a lesser extent, in Ireland) deliberately started fires to create clearings in woodland, where succulent new growth attracted grazers. By providing more food, this practice also fostered an increase in deer numbers and created healthier and fatter animals – a sort of 'proto-pastoralist herd management'.[20] People may also have gathered fodder to attract deer to places where hunters lay in wait, and selective culling may have been employed to remove surplus stags, leaving more resources for pregnant does.[21] It has also been suggested that people may have created safe spaces where deer were not hunted, so that they could rear their young in peace.[22] Such practices would have affected the ecosystem in complex ways, especially the wolves who preyed upon and competed with humans for these deer.

Mesolithic peoples also maintained intimate, spiritual relationships with the environment around them. An important Mesolithic site in the East Riding of Yorkshire, known as Star Carr, provides extensive evidence of this. First excavated in the middle of the twentieth century, Star Carr is now composed entirely of farmland, though it used to sit on the edge of a large body of water known as Lake Flixton. By 11,000 years ago this lake was slowly beginning to be enveloped by peat, but it was still being utilised by a group of Mesolithic hunter-gatherers who left behind thousands of artefacts. This site is remarkable for numerous reasons, not least the presence of a dwelling often described as the 'oldest house in Britain', and a wooden platform at the lake's edge which provides the earliest evidence of carpentry ever found in Britain.[23]

Also present were over a thousand bones from the animals that the people of Star Carr had hunted, eaten, and utilised the body-parts of, including hundreds of antlers and bones belonging to red deer (which were much larger than their modern ancestors),[24] as well as the remains of roe deer, elk, aurochs, wild boar, hedgehog, wildcat, beaver, hare, pine marten, badger, bear, and red fox, along with various bird and fish species.[25] Jewellery made from the teeth and bones of these animals, as well as from amber and rocks, was also present. There were numerous tools with which the people killed their prey and processed their bodies, and there was even part of what seemed to be a small paddle, which may have been used to steer a boat that transported the kills.[26] Perhaps most remarkable, however, are the 21 headdresses constructed from the smoothed-down front halves of red deer skulls, antlers included. These eerie masks, bored with eye holes to allow the wearer to see out, may have functioned as a sort of disguise when hunting. An alternative, less utilitarian purpose may have seen them used in ritual dances, or even to facilitate a 'transformation' in which the wearer harnessed the deer's characteristics.[27] These headdresses are exceptionally rare, as are the numerous antlers found at Star Carr that have been sharpened to points and carved with barbs, which may have been used to hunt the very animals from which they came.[28]

Much of this assorted paraphernalia of prehistoric life appears to have been ritualistically placed into the lake, rather than simply being disposed of as rubbish in the peat. The trove of red deer antlers discovered here appear to have been 'saved' throughout the year and brought to Star Carr from all over the surrounding area to be deposited in the peat, a reflection of the complex perceptions of and relationships with animals held by people in Mesolithic Britain.[29] Likewise, intact limbs, skulls, and potentially even whole, unbutchered corpses of red deer and elk were put into the bog, seemingly as part of a complex, possibly animistic belief system.[30]

Among the thousands of animal remains unearthed at Star Carr were two dog skulls and some leg bones, as well as an almost entirely intact dog skeleton deposited in the marsh around 11,000 years ago. These remains may have been placed in the lake in acknowledgement of the protection and companionship that these animals provided.[31] The metatarsal of a wolf which may have once been attached to a pelt was also found in the lake.[32] But despite belonging to the decidedly undomesticated counterpart to the dog whose bones were interred in the lake, this wolf's remains were treated in a similar way. Perhaps they were the subject of a ritual in which the people who killed the wolf repaid it for 'giving [itself] up to the hunter',[33] suggesting a deep relationship with the species that went beyond simple exploitation for its fur or other body-parts.[34]

The practice of treating wolf remains in special ways was not unique to Star Carr. A canid rib fragment uncovered at a late glacial or early Mesolithic site in Uxbridge had been polished to give it a unique glimmer which set it apart from every other object found in this assemblage, suggesting that this piece was selected, worked, and preserved for a specific reason. Similarly, about 40 miles away at another Mesolithic site in Berkshire, two wolf jawbones had been worked by people before being deposited in the earth.[35]

The Mesolithic people of Britain evidently did not have a merely utilitarian relationship with and keen knowledge of the natural world around them. They had deep, meaningful connections with the landscape's contours and valleys, its topographical and floral maps and the animals within them, not just on a species-by-species basis but even, perhaps, with particular groups or individuals.[36] For these hunter-gatherers, animals seemingly represented not only sources of food and other raw materials but beings with agency, souls, and spiritual lives of their own, who made the ultimate sacrifice to keep people fed, clothed, and warm. Accordingly, they were afforded a level of respect which appears to have affected how their bodies and souls were treated in death.[37]

Though it is likely that these people encountered the elusive wolves with which they shared the landscape only infrequently, they will often have heard ethereal howls drifting over the treetops. An alien yet familiar species who looked and behaved both like and unlike dogs, wolves were also strange mirrors of humans. Wolves and humans shared the same 'rhythms' – both species followed herds of deer on the move and hunted cooperatively in order to bring down their prey, the last of the native carnivores to do so.[38]

Their uncanniness ensured that rare meetings with wolves will have left a mark upon any humans who encountered them. Perhaps their bones were deliberately selected, worked, and specially deposited in acknowledgement of their 'otherness', their uniqueness, their rarity, or as a reminder of a particularly influential encounter. These artefacts may also have functioned as amulets believed to endow the wearer with the wolf's strength, speed, and fierce determination, allowing them to see the world through lupine eyes and mimic the wolf's movements, particularly when hunting, or even to 'transform' into a wolf.[39]

The Dawn of the Dog

The close affinity between wolves and humans, as well as the wolf's sociability, may lie behind the former's domestication. These animals slotted so neatly into human society because, in many ways, it was similar to the wolf's own.[40]

It is unclear where, when, or how wolves transformed into dogs.[41] Some theorise that wolves were the instigators of this novel arrangement. Attracted by the wide array of waste materials produced by human communities, it is thought that wolves who were less fearful of people began to scavenge from them. Initially they made forays into human company only infrequently. But eventually they became firm fixtures of human camps as synanthropes (species who have adapted to benefit from living in human environments, like the modern raccoon), where they were tolerated by humans since they provided a free waste-disposal service. The boldness of these wolves was rewarded by a meal, and natural selection soon began its work. This advantageous trait was passed down through the generations, ultimately leading these scavenging specialists to diverge from the wolves who remained wild. In time, the humans recognised that these new, tame campmates could be useful assets, not only because they could lead them to and help bring down prey, but also because they proved useful guards against incursions from other people or animals. These people may, therefore, have favoured the individual animals that proved most useful to them and who were the most sociable, resulting in artificial selection which ultimately gave rise to the domestic dog.[42]

An alternative theory suggests that late Pleistocene and early Mesolithic hunter-gatherers may have captured and raised wolf pups. A twentieth-century study on selective breeding to encourage friendliness in silver foxes offers insight into how this may have happened. While the silver fox is notoriously aggressive, selective breeding of individuals who responded least negatively to human contact led to the sixth generation of foxes not only accepting human attention but actively seeking it out. They licked people's hands, enjoyed being handled and petted, wagged their tails, and even whined when humans left them. Remarkably, the changes were not only behavioural but physiological. The foxes' ears became floppier, their tails became curly, their coats became patterned, and their snouts became less pronounced (all hallmarks of domestication across multiple species), despite none of these traits having been deliberately selected for through breeding.[43] This remarkable result demonstrates just how quickly wolf domestication could have happened if it took place through the capture, socialisation, and selective breeding of wolf pups who were the most receptive to human company. Within just a few generations, late Pleistocene humans could have created a clade of proto-dogs that were both psychologically and physiologically distinct from the wolves of the wild.

But why should humans take such a risk, deliberately bringing a potentially deadly animal into their midst and inviting the anger and desperation of a frantic wolf mother? There are many ideas, though the most widely accepted is that they were intended to help with hunting. Anthropologist Pat Shipman has suggested that the impetus behind the initial taming of wolves was 'to create "living tools" so that humans could borrow the useful attributes of the domesticates that were not shared by humans', such as endurance, scent-tracking, and harrying prey, a collaboration which made the hunt easier, less risky, and more successful and bountiful for both species.[44] They may also have been used as guard dogs, protecting both encampments (and perhaps those left behind while others went to hunt) and prey carcasses against other wild animals, as well as for transportation, pest control, disposal of waste material, as 'living weapons' during conflicts, as 'bed-warmers' and, after the advent of agriculture, for herding livestock.[45] It has been suggested that the Star Carr wolf was employed in herding even before the domestication of sheep, to 'drive a flock or herd of animals towards concealed hunters'.[46] If desperation called, dogs could also serve as a useful source of food, fur, and bone for crafting weapons and tools.[47]

Like all other living beings, dogs affected the ecosystems they were part of.[48] Not only did they eat the food they were given by humans, which they may have helped acquire, but they may have gone on their own jaunts into the woods, honing their hunting skills on small mammals and birds.[49] The presence of dogs would have particularly strongly affected their wild cousins. Having these proto-dogs around would have made it even more difficult for wolves to compete with humans for resources and space. As a territorial species, wolves are likely to have attacked or, at the least, harassed the early domesticated dogs, creating greater tension and competition between wolves and humans than there had been previously. In an effort to protect their canine companions, Mesolithic people may have killed wolves.[50]

On the rare occasions when wolves and dogs did not have a negative interaction they may have got a little too friendly, resulting in hybridised offspring.[51] It has been suggested that an approximately 5,500-year-old canid skull found in Surrey may have belonged to one such wolfdog. The large size of this skull and the spacing of its teeth are more typical of a wolf (dogs' teeth tend to be crammed closer together than wolf teeth, a product of domestication which saw their teeth shrink at a slower rate than their jaws), but the teeth themselves, being smaller, are more dog-like. Also suggesting that this was a domesticated animal is the fact that it had suffered two significant blows to its head which had distorted the shape of its skull, injuries unlikely to have been survived by an animal living in the wild. While head trauma is evident in many dog remains, these specimens usually date to the Roman period or later. It has therefore been suggested that the injuries sustained by this particular animal may be the result of efforts to control a wolf–dog hybrid; such animals are bolder than wild wolves and can be very aggressive and unpredictable, making them particularly dangerous to keep as pets and far harder to control than dogs. This canid could be the product of unintentional interbreeding caused by increasing encroachment of human settlements into wolf habitats (which may have been modified or shrinking), or by falling numbers of natural prey leading wolves to venture closer to human communities. It could even have been the result of deliberate interbreeding orchestrated by humans in order to produce larger animals with greater strength than the domesticated dog.[52]

With their new companions in tow, humans occupied an even more dominant position in the natural world than before. Humans had not only developed the ability to use tools and weapons to their advantage – now they were faster, they possessed a better sense of smell and better eyesight, and they had greater endurance and stamina. And they did not even have to evolve these traits themselves – instead, they borrowed them from another species.[53] Humans were fast becoming powerful agents of change in the landscape: a transformation which, in the following era, was to become even more dramatic as a revolutionary invention from the Near East swept across Europe.

From Friend to Foe: The Agricultural Revolution

One summer five or six thousand years ago, a wolf (or, perhaps, a wolf's domesticated cousin) trotted its way up the coast just north of what is now Liverpool, leaving squelchy pawprints in the mud of a marshy reedbed as it went. Baked by the sun and subsequently protected beneath dunes, this animal's prints – along with those of deer, aurochs, horse, wild boar, a variety of bird species, and humans (the latter so detailed we can see the whorls, ridges and spirals of their toeprints) – were preserved until the coastline changed and the sand was blown away some thousands of years later. These pawprints have an immediacy that belies their age. Modern-day observers of these ghostly imprints would be forgiven for looking into the distance to seek the swishing tail and retreating silhouette of the individual that left them.

This animal belonged to a changing landscape. Around 6,000 years ago, Britain and Ireland saw the arrival of a discovery that was already leaving an indelible mark upon the world, altering the course of human and natural history alike.

Farming.

The Agricultural Revolution that eventually reached the shores of Britain and Ireland began much farther afield, in a bow-shaped area of the Near East known as the Fertile Crescent (in what is now Iraq, Syria, Lebanon, Palestine, Israel, Jordan, Egypt, Kuwait, Turkey, Iran, and Cyprus). Here, around 11,500 years ago, hunter-gatherers began to cultivate crops and, apparently at around the same time, to manage domestic stocks of goats, mouflons, aurochs, and wild boars for their meat.[54]

How and why the domestication of plants and livestock first came about is unknown, particularly since it is a more labour-intensive way of life than hunting and gathering. Some theorise that a warming climate made wild food sources less reliable and crop growing easier, while others suggest that human populations were increasing so much that hunting and gathering was no longer viable. Overhunting and overexploitation of resources is another suggestion, a theory supported by archaeological evidence suggesting that people living in the Near East shortly before the advent of farming were eating smaller and less easily caught animals, a necessity since larger prey species had all but disappeared.[55] Or it could be the opposite, with an abundance of resources allowing people to stay in one place and affording them surplus energy and time to experiment with farming.

It has been suggested that the domestication of wolves could, ironically, have presaged the very thing that ultimately led to their demise over so much of the world: the domestication of livestock. Perhaps the taming of wolves planted the seed of an idea – an idea that would free hunter-gatherers from the whims of the natural world and the wild animals upon which they relied for their meat. The inspiration may even have been drawn from the natural behaviour of wolves 'herding' ungulates during the hunt. Perhaps prehistoric humans were inspired to use this hunting tactic to control and subdue herds of prey animals, transforming 'a feral food resource into a controlled flock' as author Sally Coulthard puts it.[56] Prey animals subject to this practice may soon have developed a flocking instinct, ultimately leading to their domestication.[57]

Whatever its origins, agriculture spread from the Fertile Crescent, whether by the physical movement of domestic stock or else via the sharing of knowledge. Farming made its slow but inevitable way westwards until, around 6,000 years ago, domesticated plants and goats, sheep, cattle, and pigs (or the idea of them) had completed the 2,000 mile-journey from the Fertile Crescent to the far western reaches of Europe.

Curiously, farming appears to have reached Ireland and the Isle of Man before it came to mainland Britain,[58] though it spread quickly once it arrived in the latter, taking only a few centuries to disseminate throughout the island.[59] Although people in Britain and Ireland were likely already manipulating their environment to procure food (such as through controlled burns to attract grazers and perhaps 'farming' hazelnuts by pruning their shrubs to increase yields),[60] and hunting and gathering continued alongside farming (which initially may have involved keeping just a few livestock animals),[61] practices slowly but surely changed.

The impact was profound, affecting people, wild animals, the landscape and, of course, the new domesticated inhabitants themselves. For the first time, humans could put down roots. They were no longer bound by the movements of wild prey but able to rely almost entirely on cows, sheep, pigs, and goats for both meat and, later, clothing.[62] Domestic animals also provided milk (a valuable source of protein), leather, useful materials such as horn, and fertiliser for crops. They could also be used to complete otherwise gruelling tasks such as ploughing, making crop-farming less labour intensive.

The human population grew, their new lifestyles both allowing and requiring them to have more children to share the labour.[63] With a reliable supply of food wherever they ventured, they spread far and wide into previously uninhabited areas.[64] Communities expanded, houses were built, boundaries were demarcated and, eventually, field systems were put into place. More and more objects were produced to keep up with the demands of this new way of life, requiring the exploitation of ever more resources.[65]

Much of the forest which had characterised both Britain and Ireland was cleared and burned to make room for settlements, crop-planting, and pasture, as well as to provide fuel and building materials. More and more woodland disappeared as the Neolithic progressed and humans became ever more prolific. Domestic animals in turn sped woodland clearance along, with cows, sheep, and goats happily browsing from the shoots of felled and burnt trees.[66] These huge changes in the landscape were good news for grassland-, heathland-, and downland-dwelling animals such as field voles, but bad news for forest creatures like pine martens and wildcats.[67]

But although docile domestic animals made food much more accessible and provided greater security than living off the land, looking after them was time consuming. They needed to be herded, put out to pasture, provided with food, guarded, and prevented from eating crops. Later, when sheep began to be bred for their wool, even more labour and even more sheep were involved. Plucking, sorting, washing, processing, spinning, and weaving meant that fabric could take over 100 days to produce.[68] Livestock were valuable investments into which people poured their time and energy, upon which their livelihoods – both present and future – depended. Whether to sudden changes in climate, epidemics, or the predators who suddenly had a very enticing larder on their doorstep, the loss of these animals was devastating.

Wolves were no longer mere competitors with whom humans had more or less peaceably shared the landscape for thousands of years, animals which they had tried to emulate and embody, perhaps even admired, the first creatures with which humans had struck up an alliance. Now, they were trespassers and thieves, objects of fear and hatred, perhaps even seen as bloodthirsty and greedy.[69] There was no room for empathy or identification with the animals who threatened the food security and survival that was staked upon domestic animals.[70] Wolves were driven away from livestock pasturage and human settlements, and those who did not stay away were probably hunted down.

The Neolithic Agricultural Revolution was undoubtedly one of the most significant events in the history of wolf–human relationships, not only in Britain and Ireland but worldwide. It was part of a larger redrawing of humankind's relationship with the natural world, which saw humans begin to shape the landscape around them rather than being shaped by it. Wild animals had little choice: adapt to this new world dominated by humans and their domesticated animals, or perish. For humans, meanwhile, as they grew further and further apart from the rest of the natural world, wild creatures increasingly became more and more alien, more and more 'other'.

In spite of – or perhaps because of – their otherness, wild animals frequently featured in the spiritual lives and deaths of Neolithic humans. Wolf remains are often found in funerary contexts across Britain and Ireland. In Wales, for example, the bones of three adults and one child were placed alongside various artefacts and animal remains in a cavern on Great Orme (the limestone headland that overlooks Llandudno), including the decorated and perforated canine tooth of a wolf.[71] A cavern in the Edenvale-Newhall cave complex in Co. Clare also housed Neolithic wolf teeth along with human, red deer, and bear remains. One such wolf tooth was stripped of the enamel and etched with a groove, while another was perforated so that it could be worn as a necklace.[72]

When worn as pendants, these teeth perhaps functioned as amulets to protect the wearer or to allow them to acquire desirable lupine characteristics, such as ferocity and courage.[73] It has also been suggested that such artefacts may have belonged to the hunters who had killed the animal from which the tooth was extracted, likely in response to or to prevent livestock depredation. Yet although hunters played a valuable role in ensuring the security of their communities, potentially controlling predators or providing resources to supplement the diet, such people may have occupied a liminal position in the new societies centred around agriculture. They were practitioners of the 'old' arts, which required them to channel the hunting prowess of the animals they targeted – the very same animals that had become rivals and direct threats to human livelihood. Thus, hunters were perhaps seen as 'wild' or 'animalistic' in some way. The wolf tooth pendants may have emblematised their wolfishness or helped them to channel a lupine nature – a microcosmic artefact in which the essence of the wolf was contained, whose traits could be transferred onto its possessor. The cave burial sites in which these items have been found may be significant in this context. Perhaps the hunters were placed in the home of the cave-dwelling wolf to be among kindred spirits, or perhaps they were feared for their animality and were therefore buried in caves to keep them far away from settlements.[74]

Despite their inimical status, wolves may also have been totem animals for certain communities, their remains used in rituals or serving a spiritual function. A roughly 5,500-year-old wolf leg bone found in a Yorkshire fissure may have functioned as a talisman of sorts, especially given that it was discovered alongside the smashed bones of cows and a collection of human skulls which may have been deposited with their flesh still intact.[75] The atlas bone of a very large wolf was likewise found buried with human remains at Stonehenge alongside the bones of foxes and ravens,[76] while wolf, sheep, and cattle remains that are around 4,500–5,000 years old have also been unearthed at Orkney's Stones of Stenness, which may have belonged to animals sacrificed at this centre of ritual.[77]

Such practices continued into the Bronze Age, an era that began around 4,500 years ago with a boom in the production and trade of the alloyed metal for which the period is named. The Bronze Age saw the intensification of farming practices, with much of the land parcelled up into fields during this period. Archaeological investigations in eastern England have uncovered evidence of very large flocks of sheep, but a dearth of evidence for the hunting of wild animals.[78] But despite the threat they posed to sheep, wolves still continued to appear in funerary contexts. Perforated teeth have been found in both cairns and caves in Co. Clare, one of which had been polished.[79] This polished tooth is one of many artefacts placed in a cave as part of a ritual, alongside offerings including pottery; beads made from shells, bone and amber; tools; human bones; and the remains of juvenile animals which had apparently been sacrificed.[80] Similarly, a 4,000-year-old urn containing the remains of two adults and three children was found buried alongside the burnt remains of a wolf or large dog in Tipperary, while a swallowhole in Fermanagh contained wolf or dog bones alongside the remains of four Bronze Age humans, cattle, sheep, pigs, red deer, and hares, which may have been thrown in as part of a ritual.[81]

By this time, wild species were less commonly exploited than domesticated animals. In light of this, it has been suggested that their body-parts had greater symbolic appeal and meaning, a conclusion supported by the fact that red deer antlers from this period – both worked and unworked – are encountered more frequently in funerary contexts than at settlements. Wolves, in particular, as dangerous predators who were closely linked with caves,[82] may have been associated with death or considered guardians of an underworld represented by or entered through the shadowy caverns in which human bodies were sometimes interred.[83]

People since the Neolithic may even have deliberately allowed wolves (as well as foxes and even dogs) to scavenge from human corpses, in a process called excarnation. An early Neolithic funerary barrow discovered in a Cotswold village contained the remains of several humans who had been subject to this ritual. Their bodies were likely left outside to be de-fleshed by wolves before being emplaced in the enclosed tomb, a funerary rite apparently closely watched by someone who made sure the bones were cleaned but not damaged.[84] A Welsh Bronze Age burial mound in Denbighshire also contained human bones that appear to have been treated in the same way – ten of these bones had been picked clean and divested of their marrow by a carnivore with a powerful jaw, most likely a wolf, while smaller bone fragments were seemingly gnawed by juvenile wolves. After the wolves had reduced the body to a disarticulated skeleton the people returned to claim the bones, carefully positioning them in the cave before covering them with stones.[85] A number of 'excarnation structures' – raised platforms onto which the dead were placed – have also been found throughout Britain, from Orkney to Bedfordshire, often next to or incorporated into burial monuments.[86]

It is possible that this ritual was almost sacrificial in nature, an act of appeasement or propitiation to prevent wolves from preying on livestock.[87] Or maybe it was part of a complex network of beliefs, a 'vital exchange of matter' and 'a necessary process for maintaining the flow of human, animal and spirit life – a cosmic food web', as archaeologist Joshua Pollard puts it.[88] It is perhaps because some Neolithic and Bronze Age people left their loved ones to return to the earth (or journey to the afterlife) via the stomachs of wolves that these animals appear so infrequently in archaeological records from this time. Hunting such animals may have been considered taboo in light of their role in the 'transformative processes' of excarnation, their eating of the dead 'ascrib[ing] them a status as "impure", dangerous or powerful beings'. Consuming wolves was perhaps considered 'tantamount to cannibalism'.[89]

In some places, however, there may have been attempts to protect human remains from the depredations of wild animals. In a cave burial in Co. Limerick, the bones of one or two Neolithic-age humans were accompanied by a dog buried underneath stones at the entrance to the cave. This canine could have served as a symbolic deterrent to wild animals attempting to cross the threshold and scavenge the remains within, in the same way as physical barriers – slabs of stone to block access, for example – were sometimes used to keep out marauding animals, including wolves.[90]

Celtic Cosmic Wolves Meet Romulus, Remus and the Romans

The spread of Celtic culture throughout Britain and Ireland during the Iron Age (c.2,800–1,900 years ago) saw the arrival of new ideas about wolves.[91] Celtic spiritual beliefs were animistic – every part of nature was thought to possess a spirit. Gods were represented in elements of the natural world, and animals could be seen as divine in and of themselves, as messengers of the gods, or as gods in animal or part-animal form.[92]

Life was still largely rural. Ever greater swathes of land were subsumed by farming, the landscape divided into fields and split by roadways. Whether a cause or consequence of an increasing population, agricultural practices were becoming more intensive.[93] Livestock remained an important part of local economies and livelihoods. The Celts on the Continent were apparently so sheep-savvy that first-century CE geographer Strabo commented on their 'enormous flocks of sheep and herds of swine'. Contemporary records suggest that little was different across the North Sea, where the Celts in Wales and Ireland were likewise heavily reliant on cattle, sheep, and pigs.[94]

Protecting these vital assets was paramount. Ringforts and hillforts (the latter a type of settlement found atop a natural elevation, surrounded by ditches and fencing mapped onto the hill's contours, while the former was protected by ditches) were common throughout Britain and Ireland by the first century CE. These forts served as enclosures to protect flocks and herds against thieves and predators, who may also have been deterred with dogs and slingshots.[95] Such forts are particularly prevalent in Ireland, where it is estimated that tens of thousands of such structures once existed. This is perhaps evidence of an emphasis on 'reducing the negative impacts of wolf predation [...], rather than on wolf elimination',[96] possibly a result of animistic spiritual beliefs about the natural world which survived longer in Ireland than in Britain.

Advances in metalworking during this time facilitated improvements in farming as well as clothing production. Garments fashioned from wound fibres became more common than clothing made from pelts.[97] Since wild animals were infrequently exploited for this purpose and very rarely consumed, hunting became even more obsolete. Wolves still had a few uses, however, such that some appear to have been imported to the northern Scottish islands around this time – a possible wolf skull found on Orkney may have arrived attached to a pelt, while a wolf jaw uncovered on Shetland was imported to the island for some reason, likely from either mainland Scotland or Norway.[98] Very few other wolf remains survive from this period, possibly due to their waning utility but also because domestic animals were sacrificed to the gods far more often than wild animals, which were perhaps eschewed because they were considered too closely connected with the divine, or possibly too akin to humans.[99]

The arrival of the Romans in Britain heralded significant cultural changes. Though they were unable to push further than southern Scotland nor claim any land across the Irish Sea, the Romans succeeded in conquering all of what is now England and Wales, which they occupied between 43 and 410 CE. Roman and Celtic cultures collided, gradually creating a distinct Romano-British civilization.

Although farming was already well established by the time of their arrival (the greatest change brought about by the Romans was instead the creation of urban environments),[100] the Romans did expand and advance existing agricultural systems, which improved productivity. By the second and third centuries CE large estates owned by the upper classes were beginning to appear in Roman Britain, and Roman statesmen and emperors sent money to the owners of villas to develop native agricultural processes. The Empire perhaps even sponsored deforestation, as well as the draining and settlement of fenland. Livestock farming was increasingly intensive. Cows were the preferred source of meat, and some settlements housed vast herds of animals larger than any seen before. Some sheep varieties were bred specifically for their wool, a significant industry in the later years of the Roman Empire, when entire settlements in southeast England were potentially devoted to raising these animals.[101] Influenced by contact with Roman Britain, the Celts in Ireland appear to have followed suit, clearing forests and expanding farming in the third and fourth centuries.[102]

More than 3.5 million people called Roman Britain home by the turn of the fourth century CE, though only a quarter of a million of them lived in urban settings.[103] Settlements tended to be concentrated in the lowlands of the south and southeast, with less of a mark made on the uplands of the north and northwest.[104] Much land in the south was taken up by villas and their attached estates, which were interspersed with small towns and farmland. Woodland was cleared to create space as well as to provision house- and ship-building, to fuel baths, and to create metals and glass. What few forests remained – and in some areas, there were none left at all – were found in small patches between farmland.[105] Reclaimed fenland also housed a burgeoning population by the second century CE, and may have been used for grazing livestock in the warmer months.[106] The Romans may even have created enclosed hunting parks modelled after those commonly found in Italy and Gaul which, according to first-century BCE Roman writer Columella, housed both native deer as well as exotic species such as fallow deer and gazelles.[107]

Despite their reputation as prolific livestock farmers, it is unclear whether the Celts hunted wolves in any great numbers. Medieval Scottish historian Hector Boece (1465–1536) imagines legendary first-century BCE Scottish king Ederus as an enjoyer of wolf hunting, which he was said to pursue with the help of hounds. Wolves were a decided nuisance in this age, according to Boece; the historian details how another legendary king named Dorvadilla decreed that anyone who killed a wolf (which were supposedly hunted enthusiastically at this time because they were 'very harmful to cattle') should receive a cow in return.[108] Aside from a necklace made from a wolf tooth found in Yorkshire, there is little archaeological evidence to support or deny these medieval myths.[109] However, it is possible that the meagre record is not representative of the true nature of wolf–human relationships at this time – perhaps the kills were simply processed where they fell, rather than in the settlements that have been excavated.[110]

Wolves may have served as a totem animal for the Iceni tribe, who occupied part of what is now East Anglia around the first century BCE, and whose leader, Boudicca, is famous for resisting Roman rule. Golden Icenian coins (Figure 2.1) depict a crude wolf, bristling with fur – perhaps in aggression – and with mouth agape, in a seemingly very deliberate move away from more traditional equine coinage decoration.[111] It may be that this lupine emblem, which is otherwise rare on Celtic coins from Britain, was a symbol of Iceni independence, their wild spirits that would not be quashed by Roman invaders.[112] Indeed, several centuries later the Roman author Cassius Dio depicts Boudicca describing her people as 'dogs and wolves', over whom the Roman 'hares and foxes' had no right to rule. The Iceni might have admired or identified with the wolf, much like a European Celtic tribe named the Volcae, whose name may be derived from the Proto-Celtic word for 'wolf'. The Volcae appear to have lent their moniker to an early Germanic term for foreign people, *walhaz*, which was used of the Romano-Celts and eventually became the country name we know today: Wales.[113]

Rather than speaking to specific cultural identities, however, these numismatic wolves may instead tell us of the spiritual and cultic beliefs of some Celtic people at this time, particularly since such coins may not have been used for trade but as tokens for sacred offerings to the gods. One scholar reads these coins as a narrative depicting sun- and moon-eating cosmic wolves, much like the wolf of doom in Old Norse mythology, Fenrir, who swallows the sun during the apocalyptic event known as Ragnarǫk. According to this interpretation, the Icenian wolf on one coin type is a sun-chaser, creeping across a bronze fenland as it stalks the divine sun on its upwards journey (the name of the Norse wolf Fenrir, incidentally, means 'fen-dweller'), while the

Figure 2.1. Icenian coins depicting wolves. © Chris Rudd Ltd. www.celticcoins.com

Figure 2.2. Figurine found at Woodeaton, depicting a possible wolf-god.

wolf on another Icenian variant pursues a crescent moon (or a moon about to be eclipsed) in the opposite direction.[114]

A small figurine unearthed in the Oxfordshire village of Woodeaton (Figure 2.2) also appears to portray a lupine god. This extraordinary carving depicts a sitting wolf from whose mouth protrudes part of a much smaller human body which has been devoured head first. Dating from sometime between the first and fifth centuries CE, this artefact almost certainly had some now-lost religious meaning, since it was found at the site of a Romano-Celtic temple. The statuette could have Celtic or Roman significance, perhaps in relation to Mars,[115] the Roman god of war who is often associated with wolves, or else a Celtic deity linked to the animal.[116]

Another figurine that may represent a wolf god has also been found in Devizes, Wiltshire, though this animal is in motion rather than seated, and the gruesome body hanging from the mouth is here replaced with only a lolling tongue.[117] Two other statuettes found in Conwy feature a similar protruding tongue, and may also represent a canine or lupine deity of sorts, perhaps one associated with death or the underworld.[118] Likewise, the man consumed by the wolf of Woodeaton perhaps represents a person undertaking a journey in death to meet a lupine underworld god or a totemic ancestral spirit.[119] A wolf head also appears between the legs of a bear carved from stone found in an Iron Age burial ground in Co. Armagh, which may have been used in burial rituals or as a votive offering to ancestral spirits or gods.[120]

These artefacts suggest that wolves played an important role in Celtic cosmology in Britain and Ireland. Indeed, we know that British Celts did worship a god called Cunomaglus, whose name may be indicative of a canine or lupine nature. *Cuno* meant both 'wolf' and 'dog', and paired with *maglus* this eponym meant 'lord of the wolves' or 'lord of the dogs'. A lupine meaning may be more likely, however, since this deity is described as 'the god Apollo Cunomaglos' on a Roman-age shrine in Nettleton Shrub, Wiltshire. Apollo was a Greek god closely associated with wolves, described as 'wolf-born' since his goddess mother transformed into a she-wolf to birth him. His mother's lupine form during his entrance to the world apparently left its mark on Apollo, who sometimes used his wolfishness to help defeat enemy armies, while wolves also served as his guardians and messengers. The pairing of this deity with Cunomaglus may suggest perceived similarities or overlap between the two, indicating that the Celtic god may too have been wolfish. Moreover, large numbers of sheep and/or goat bones, many from young animals, were also found at Nettleton, corresponding with a mention in eighth century BCE Greek poet Homer's *Iliad* that lambs were sacrificed to the 'wolf-born god' Apollo. The specific role that Cunomaglus played in the spiritual beliefs of those who visited this shrine and perhaps others elsewhere in Romano-Celtic Britain is unknown, though it is possible that, like Apollo, with his defensive wolfishness and his guardian messenger wolves, Cunomaglus may have served a protective function for those who worshipped him. Paradoxically, this lupine god may even have functioned as a protector for farmers, keeping deer and boar at bay from the crops that were so important to the local economy of the Cotswolds at this time.[121]

The wolf also featured frequently in Roman myths and legends. The most important and best known in Britain was the tale of Romulus and Remus who, so the story goes, were suckled by a wolf after being abandoned in the woods as infants. Romulus later killed his brother and founded the city of Rome. The she-wolf thus became a symbol of Rome itself and of the empire's strength and power, as well as an expression of loyalty to it.

The she-wolf and twins are depicted on a variety of objects from Roman Britain, particularly on coins issued by the Roman Empire. Fourth-century CE coins of Constantine the Great featuring the wolf and twins have been found in England, while coins with the motif issued under Domitian Caesar in the first century CE have been uncovered in Wales and England. The wolf and twins are also depicted on a sword scabbard found in the Thames at Fulham, possibly intended to protect the person who wielded it, or to impress anyone who saw it.[122] A second-century CE bronze figurine from Wiltshire features a charming she-wolf with the twins below, as do a panel from a shrine in Corbridge,[123] and a first- or second-century CE bronze mount found in London which may originally have been used to decorate furniture.[124] But by far the most memorable example of the motif is on a third- or fourth-century mosaic preserved in Aldborough, Yorkshire (Figure 2.3), though it is noteworthy for unfortunate reasons. Here, an enormous black she-wolf grins vacantly at the viewer while two tiny twins float beneath, their legs at unnatural right angles. The tree above her is at least quite charming.

Because of the key role played by the wolf in the founding of Rome, and its consequent importance in Roman mythology and identity (possibly to the extent of an almost totemic relationship, though never quite reaching levels of worship), the Romans treated this species a little differently from other predators such as lions, leopards, and bears, which were all hunted prolifically. Wolves were apparently not hunted for pleasure nor featured in the infamous battles between people and predators in Roman amphitheatres, and they seem to have been killed to protect livestock only when preventative methods – such as shepherds and guard dogs wearing spiked collars – failed. Likewise, although Roman naturalist Pliny the Elder (23/24–79 CE) describes various magical and medicinal uses for wolf body-parts in his *Natural History* – including tying wolf teeth to an infant to 'prevent [them] from being startled', or consuming a concoction of wolf liver and mulled wine to cure a cough – it is likely that he records earlier Greek and Egyptian folklore rather than practices actually carried out in ancient Rome. There is no evidence to suggest that wolves were hunted for this purpose, nor that they were hunted for any other

Figure 2.3. Roman mosaic at Aldborough, depicting a female wolf with Romulus and Remus. © Leeds Museums and Galleries, UK / Bridgeman Images

reason. Pitfall traps (the same type of traps often seen in cartoons today, which feature a hole covered by branches) are mentioned in Roman literature, but it is unlikely that they were actually deployed in practice.[125]

Wolves may have been hunted for their fur, though this was apparently infrequent. While standard-bearers in the Roman army are popularly believed to have worn wolf pelts, the evidence suggests that the animals of choice were bear and lion. Indeed, while some soldiers specialised in hunting bears or lions so that their pelts could be worn by members of the Roman army, no such equivalent role seems to have existed for wolves.[126] Only two texts reference *luparii* ('wolf hunters') at all, one of which mentions the use of poisoned meat to kill them, probably laced with aconite (also known as wolfsbane).[127]

In the rest of Roman society, cloth and wool were popular materials for clothing. Furs were worn only by rural people, labourers, and those who could not afford these more expensive woven fabrics.[128]

Even when they entered cities, wolves were simply taken as omens and safely driven back out, rather than being dealt with in safer and easier – albeit crueller – ways. Killing wolves was not a taboo, nor was it legally prohibited, but it was on the whole avoided due to the animal's special status in western Roman religion and identity. In comparison, in the eastern provinces of the empire where Greek culture was predominant, the hunting and sacrificing of wolves continued.[129] It is perhaps telling that many of the fables attributed to Aesop were not Roman but Greek in origin. Those who read these stories as children may remember the role which the wolf typically plays in such tales. Often, as in 'The Wolf and the Crane', the titular canid is a deceptive trickster. In this story, a crane bravely agrees to remove a bone stuck in a wolf's throat in exchange for a reward, but upon completing this task, the wolf declares that her reward was keeping her head. In other fables the wolf is a greedy tyrant – in 'The Wolf and the Lamb', a wolf picks a fight with an innocent lamb in order to justify killing and eating it. So too were the werewolf stories told by Roman poets like Ovid derived from earlier Greek tales, such as the myth that a Greek king, Lycaon, was transformed into a wolf by Zeus as punishment for serving the god human flesh. The choice of animal here reflects Lycaon's crime – the 'wolfishness' within him that prompted this act of savagery manifested in his lupine transformation.

But as Christianity, a religion of 'profound anthropocentrism',[130] spread throughout the Empire, Roman respect for the wolf came to a slow but definitive end. The Romans perhaps became less hesitant to hunt the animal as a result, although scenes from Christian Roman sculpture that appear to depict wolf hunting may have had more spiritual significance than resonance with real-life practice, representing the victory of faith over evil, as symbolised by the wolf.[131]

Wolves appear a handful of times in the Old and New Testaments alike, and they are not treated kindly in either. In the Old Testament the wolf is frequently associated with darkness and is used as a metaphor for ferocious and rapacious people, such as the officials of Jerusalem who are compared to 'wolves ravening the prey, to shed blood, and to destroy souls, to get dishonest gain' (Ezekiel 22:27). Wolves were considered such deadly predators, in fact, that their miraculous reversion to their Edenic vegetarian ways emblematises the peace and harmony which will be restored by Christ in the Messianic Age, when evil shall be purged from the world and 'the wolf [...] shall dwell with the lamb [...]; and a little child shall lead them' (Isaiah 11:6). This image capitalises upon the wolf's reputation as a savage beast, a stark contrast to the sheep as a symbol of purity and goodness which enhances the wondrous nature of the future kingdom of God.

In the New Testament, the wolf represents evil incarnate. Heretics and the devil are compared to wolves, who threaten righteous Christian sheep. Jesus sends his disciples into the world as 'sheep in the midst of wolves' (Matthew 10:16), while false prophets – both human and demonic – are metaphorised as 'wolves in sheep's clothing' (Matthew 7:15), preying upon the souls of God's flock whom they entrap through deceit. Christ too is a 'lamb of God', a symbol of his purity in contrast with the evil, wolfish devil, as well as a Good Shepherd who offers protection to all those who put their faith in Him. The ultimate battle between good and evil was thus encoded in the binary of the wolf and lamb.

This biblical dichotomy of wolves and sheep was drawn directly from the experiences of people living in an agricultural world.[132] It was imagery relatable for pastoral people the world over, including the prodigious sheep farmers in Britian and Ireland. They knew what a hungry wolf could do to a sheep. They could all too easily imagine the wolfish devil tearing their souls apart in the same way that the real wolf tore the organs from the bodies of helpless lambs. The immediacy of this visceral metaphor may in turn have perpetuated the conception that wolves were a physical danger to people, the threat posed by the 'wolf' to one's soul becoming entangled and mistaken with the threat of real wolves to one's body.

The Romans brought Christianity to Britain, along with these heretical and devilish wolves, around the second or third century CE. The religion slowly but surely spread, not only disseminating throughout Britain but also reaching Ireland by the turn of the fifth century. For a time, Christian and pagan beliefs were held concurrently, until a large proportion of the population of both islands followed the new religion. The wolf's natural predatory behaviour now became evidence of its savagery, bloodthirstiness, and greed. Predation was perceived as 'unnatural', not an intended part of God's creation but the product of a fallen world brought about by the original sin of the first humans.

Once admired by those Mesolithic hunter-gatherers who hoped to embody the wolf's courage, strength, speed, and hunting prowess, within a few thousand years wolves had become cunning thieves, unwanted interlopers, and bloodthirsty predators of vulnerable and valuable livestock. Now, they were emblems of the very devil itself.

CHAPTER 3

Woden and the Wolf-warriors

You are walking towards the dappled sunlight which promises the edge of the wood. There is an unpleasant smell in the air, like iron and rot. Silence hangs thickly as if draped like sheets from the trees, made louder by the absence of the clangs and cries of the battle which poured in from just beyond the trees only yesterday.

Before you reach the threshold of the forest, you hear movement. Rustling leaves, to your left. You stop, search for a blackbird fossicking for worms, or perhaps the tail of a furry forest-creature disappearing behind a tree. You see nothing. Perhaps it was the ghost of one of the unfortunate men hacked down in the battlefield beyond the trees, searching for its exit from this world into the next, to heaven or to Valhalla. You shiver.

Above: the distant scree of an eagle, its whirling descent masked by the treetops.

Edging forwards, the thinning trees reveal a prostrate figure, the first of hundreds of heaps of flesh which stretch across the bloodied and muddied field. You approach until you stand a mere foot away. A pair of eyes glares up at you, gleaming silver and vacant in the morning light. They have not yet been taken, but it is only a matter of time.

Whatever armour or weaponry he brought to the battle have already disappeared, snatched by the victors. A simple wooden cross hanging from his neck is all that remains. A gaping hole in his side where hot, throbbing organs have been pulled out betrays the visits of other scavengers. Thick, oozing blood has dried like cooling larva. Muscle and flesh have also been ripped into, an unfinished feast. The feasters will be back for more.

You wonder who this man was – what will happen to his soul which, when Judgement Day comes, will have only fragments of a body with which to reunite and rise from the dead.

Somewhere at the fringe of the forest, a sudden cronk from a lone raven startles you. A call to the feast. A great swathe of its jet-feathered friends soon appears, joining in the chorus.

In the distance you hear a reply. A howl.

Wolves in the Early Medieval World

The protracted retreat of the Romans from Britain came to its conclusion in the fifth century CE when, after many years of instability as the empire's power waned, the last Roman troops withdrew.

Throughout the final centuries of the Roman occupation, invaders had regularly arrived in Britain from the north, west, and east to test the waters. From the north came the Picts, people belonging to a group of kingdoms occupying much of Scotland who were descended from northern Celtic tribes. From across the Irish Sea came the Gaels, more Celtic descendants who occupied the numerous kingdoms into which Ireland was divided, as well as the Isle of Man and a kingdom called Dál Riata on the western Scottish coast.[1] From across the North Sea came people from a variety of Germanic tribes (primarily from the coastal areas of what is now Denmark and northwest Germany), who are today known as the 'Anglo-Saxons'.[2] Eventually, these Germanic settlers consolidated into a collection of kingdoms stretching from the northern edge of Cornwall up to the southern banks of the Firth of Forth. The Celtic Britons, meanwhile, retained control of Cornwall, Wales (which was divided into several kingdoms), and a group of kingdoms that spanned southern Scotland and northwest England, an area known today as the Hen Ogledd, or the 'Old North'.

The Roman withdrawal from Britain has traditionally been seen as a dramatic shift into a 'Dark Age' characterised by paganism, wanton violence, and cultural and intellectual decline. But this is an invention of Renaissance scholars who idealised their Roman and Greek forefathers, whose aim was to return to the 'light' of the classical period. In practice the transition was not nearly so catastrophic, nor so regressive. The political landscape certainly changed as the Romans departed and settlers from elsewhere arrived, though such changes were not sudden but a long, drawn-out process of adaptation to evolving circumstances over several centuries. What was once thought to be a seismic shift punctuated by repeated attacks and bloody massacres perpetrated by the Germanic tribes is now coming to be understood as a far more peaceful transition.[3] Likewise, the natural landscape did not simply revert to an uncultivated wilderness, nor did settlements and farms become ghost towns populated only by wild animals whose eyes glinted from the darkness of the encroaching scrub and trees.[4] Researchers now recognise that the largely rural society that had characterised Romano-Celtic Britain was little changed, with farming continuing in the same open-field system as it had for many years before.[5]

Little also changed for the wolf population. The biggest difference was the disappearance of their ursine competitors in Britain by the late Neolithic or early Bronze Age, or certainly by the early medieval period if small bear populations had managed to cling on in Scotland or northern England.[6] Bears had also disappeared from Ireland by around 3,000 years ago, the spreading human population and ever-more intensive agricultural practices shrinking their habitats.[7] It is assumed that lynx also became extinct before the early medieval period in Ireland, although very little is known of this species on the island – only a single piece of archaeological evidence attests its presence at all. The date at which this felid became extinct in Britain is a matter of debate, meanwhile. Archaeological and literary evidence suggest that it may have survived as late as the 600s in northern England,[8] and recent analysis of two texts from the seventeenth and eighteenth centuries hints at the survival of relict populations in southwest Scotland as late as 1658 or even 1760.[9] By and large, however, most wolves in early medieval Britain and Ireland did not have to compete with either of these native large carnivores.

As in Roman Britain, it appears that the wolf population was largely unexploited by people at this time. While other wild animals such as deer, boar, rabbit, fox, and a variety of mustelids are still frequently found alongside domestic species in archaeological assemblages from early medieval Britain and Ireland, suggesting exploitation for their meat and fur, wolves are extremely rare in the early medieval British and Irish archaeological record.

Although animal pelts progressively fell out of use after it became possible to weave cloth from spun fibres, clothing did feature fur trims in the early medieval period, which often functioned as a status symbol. By the ninth century there was a burgeoning fur trade in northern Europe, with evidence from Sweden and Germany suggesting that wolf furs were among the many skins shipped across the Continent at this time. But it was the soft fur of smaller animals which dominated the European fur trade.[10] In comparison to the vibrant ginger of the fox and red squirrel or the silky-soft brilliant white of the stoat's winter coat (known as ermine), wolf fur was coarse, relatively bland in colour and pattern, plentiful (trimming a cloak with ermine would take a lot more time and effort because it required multiple animals), and much less luxuriant, conferring little status. The exception is the possible use of wolf pelts by warriors, as part of ritualistic practices in which wolves were embodied or their fierce nature channelled in battle by the donning of their fur.[11]

Wolves were also probably not eaten frequently, if at all, given the ubiquity of other far more palatable sources of meat from domestic stock and wild deer. The diet of most people in England, even kings, was heavily plant-based in any case,[12] and though we might imagine opportunistic use and consumption of a wolf carcass by those who could not afford more usual (and probably tastier) sources of animal-based protein, it is highly unlikely that wolves were hunted or trapped for their flesh, especially since the consumption of wolf meat was taboo in Christian culture.[13] Even eating the marrow or meat of animals which wolves had touched was forbidden, according to both a seventh-century Irish law code and an eighth-century English penitential (a book of church rules).[14]

Though there is little evidence of the skinning or consumption of wolves, their body-parts do turn up as 'grave goods' (items which have been buried with a person), particularly in pre-Christian contexts. Canid teeth found in numerous sixth- and seventh-century graves unearthed in England appear to have been worn as pendants, judging from the holes which had been bored through them. Often found buried with women and children, these may have functioned as amulets to ward off the attacks of the species to whom the tooth belonged, perhaps in both life and death.[15]

These wolf body-parts may also have served magico-medicinal purposes. Several of the teeth found in these inhumations were kept within boxes or bags which, having been buried with women specifically, have been interpreted as the belongings of 'cunning women', healers who sometimes relied on magic. Magic and medicine were intricately connected even as Christianity took hold and the practice of interring bodies with amuletic grave goods died out. In a world without laboratories and industrially produced medicines, people had to rely on the world around them, both physical and spiritual, to provide remedies for ailments and afflictions.

As part of that world, wolves (or, rather, their body-parts) were no different.[16] In a compilation of cures known as the *Medicina de quadrupedibus* (the 'Medicine of Four-footed Animals'), an eleventh-century translation and expansion of Roman natural philosopher Pliny the Elder's *Natural History*, wolf body-parts are prescribed for the treatment of multiple ailments. For 'devil sickness' and 'ill sight' (the affliction of seeing visions), the patient was to eat wolf flesh. For those who struggled to sleep, a wolf head was to be put underneath their pillow (the practicalities of this are unclear). For ocular pain, a wolf's eye was to be pricked and placed on the sore area. After the removal of 'contrarious hairs' (presumably ingrown hairs), the marrow from a wolf's bone was to be smeared on the afflicted area to prevent them from regrowing. Even a seemingly deceased unborn child could be revived if the mother drank a concoction of equal parts 'wolfs milk', wine, and honey.[17] Another medicinal text from around the ninth century known as *Bald's Leechbook* ('leechbook' meaning 'doctor's book') also includes a treatment for a severe skin complaint in which a burnt wolf's jaw and teeth feature as ingredients, as well as a remedy for acute joint pain which required the doctor to place the ash from a burnt and ground wolf's skull on the afflicted area.[18]

While some of these treatments are more medicinal than magical in nature (though their efficacy is questionable), others are far more amuletic. The *Medicina de quadrupedibus* also details the talismanic use of a wolf tail to protect travellers against wolf attacks.[19] An eleventh- or twelfth-century charm, meanwhile, commands a 'wen' (skin-growth) to 'wither' 'under a wolf's foot' and an eagle's feather and claw, items intended to 'chase' the wen away or destroy it by channelling the ferocity and swiftness of the animals from which they came.[20]

Although the medicinal texts and grave goods show that wolves were exploited to an extent in early medieval England (although it is difficult to ascertain whether the recorded medicinal cures reflect common practices by early medieval doctors), it seems unlikely that wolves were hunted specifically for this purpose. Instead, the acquisition of wolf body-parts for magico-medicinal purposes may have been a by-product of trapping and/or hunting efforts aimed at population reduction or eradication.

Competition for prey species may have been one impetus for such efforts. The hunting of animals such as deer and boar was often a culturally motivated activity intended to show off status, power, and wealth among the elite. While the wolf offered neither physical nor symbolic benefits in this regard – there was little triumph or honour in returning from a hunt with only an animal perceived as useless vermin to show for it – the pursuit of more desirable animals like deer probably brought people into conflict with this animal's lupine predators. Although deer hunting by the nobility is more commonly associated with the Normans, who created royal hunting Forests in Britain and Ireland following the Conquest in 1066, red deer may have been hunted by the nobility in early medieval Scotland, England, and Ireland.[21] This pastime not only facilitated social interactions but also played an important role in diplomatic meetings between northern European powers.[22]

Following the example of the Frankish kingdoms, where royal Forests in which the hunting of deer was restricted to the elite were set aside from the seventh century onwards, areas comprising both woodland and clearings may have been allocated for this purpose in England.[23] By providing habitat in an increasingly deforested landscape, such places created refuges not only for deer but also for their predators, who were undoubtedly drawn to these areas due to the abundant prey that they housed.[24] Wolves which preyed on deer 'belonging' to powerful men may have been seen as competitors at best or as disrespectful thieves at worst, their predations an indirect attack on the authority of those who claimed the right to hunt this quarry.

But the greatest impetus for controlling and eradicating wolves was to protect livestock.[25] Like the Romans who came before them, the Germanic people who settled in England farmed extensively, clearing further woodland to make space for livestock.[26] Sheep were particularly important to the agricultural economy during this time (especially for their wool), and they were kept in flocks which could be as small as a few dozen but which could range into the thousands.[27] By the end of the early medieval period England's total sheep population had exceeded one million.[28] Around 25% of the landscape was used for pasture and 35% for crop growing by this stage, compared to just 15% that still comprised woodland.[29]

Cattle, oxen, pigs, and goats were also valuable, as is suggested by the fact that the Old English word for cattle (or, sometimes, any four-legged livestock animal), *feoh*, also meant 'property', 'wealth', and 'money'.[30] Sheep and goats were less valuable in Wales, although they were still kept for their wool, milk, and meat, while pigs seem to have been eaten during feasts among those of high status. But cows were the most prevalent livestock animal, functioning as a display and measure of wealth and used to value land for hundreds of years before eventually being replaced by silver in the seventh century.[31]

In Ireland, where life was almost entirely rural until the ninth century, many people continued to live within ringforts – the enclosures designed to protect livestock which were common in the Iron Age.[32] Sheep and especially cows were used as a form of currency, with wealth and status measured by these assets.[33] Cattle were also exchanged to cement social ties, beef consumption was a marker of status, and dairy was an important part of the diet.[34] Both the value of livestock in Ireland and the threat posed to them by wolves is captured in the seventh- to eighth-century Brehon Laws, which proclaimed it illegal to drive one's neighbour's livestock into known wolf territories.[35] Cows also appear to have been the most important livestock animal in Pictish Scotland, though pigs and sheep were also kept. Animal produce was an important part of the economy, with cattle possibly functioning as currency, as they did in Ireland.[36]

Given the significance of livestock to early medieval British and Irish societies and economies, depredation by wolves will have been at best an annoyance and at worst a threat to one's livelihood. But it is unclear how often depredations may have occurred. The image of wolves preying on sheep was a culturally common one in Christian communities, but wolves tend to eschew domestic livestock so long as there is enough wild prey available, and there are few actual records of depredations from this period.[37] Since deer were probably hunted primarily for sport rather than for their meat, it seems likely that there was sufficient wild prey to dissuade wolves from attempting to take livestock very often, especially given that domestic animals were apparently guarded against such incursions, making them a less attractive food source[38] – one tenth- to eleventh-century English text features a shepherd who describes how he 'drive[s] [his] sheep to their pasture and stand[s] over them with dogs in the heat and in the cold, lest wolves swallow them up'.[39] Sheep were also kept in enclosures to protect them from wolves, with numerous English charters referring to such places.[40]

Similarly, livestock in Scotland appear to have been kept indoors, protecting them from both predators and the elements, a practice that may have become more common as time went on.[41] But perhaps the most unique form of protection was seen in Ireland, where bulls appear to have been used to defend cattle herds from wolf depredations. A poem recorded in writing in the twelfth century (but which possibly circulated by word-of-mouth prior to this) mentions a 'brindled bull' protecting its herd from a host of dangers including 'wolf-packs'.[42] The Brehon laws also mention a *dam conchaid*, which possibly refers to a 'wolf-fighting ox'.[43]

It is perhaps due to their depredations upon domestic animals that the hunting of wolves was legally mandated in early medieval Ireland.[44] The Brehon Laws dictated that the tenants of a lord were required to hunt a variety of undesirable beings including pirates, members of other tribes, and wolves, the latter once a week.[45] A similar law required a person to come to their lord's aid to defend against 'pirates, robbers, and wolves'.[46]

It is frequently said that wolves were hunted to extinction in Wales in the tenth century, although the veracity of the story from which this claim originates is debatable. According to this tale, the English king Edgar (r.959–75) demanded an annual tribute of 300 wolves after defeating the Welsh king Idwal in battle. Idwal faithfully paid this tribute until, in the fourth year, he was forced to default because he could find no more wolves to kill. This account is first recorded in the *History of the Kings of England* by twelfth-century historian William of Malmesbury,[47] written around 1124. The account was expanded upon by English clergyman William Harrison (1534–1593) in his contribution to *Holinshed's Chronicles of England, Scotland, and Ireland*, in which he claimed that the corpses 'were buried at Wolfpit in Cambridgeshire'.[48] This addition seems to be little more than an imaginative fancy, an interesting detail to neatly tie the legendary tribute to a place with a fitting name, or to explain the name itself. It is testament to the story's enduring appeal that numerous legends have continued to spring up around this tale. Victorian antiquarian Charles

Bohun Smyth noted that 'two wooden wolves' heads still remain near Glastonbury on an ancient house, where [...] King Eadger lived and received annually his tax from the Welsh in 300 heads'.[49] An iron wolf on a door in a twelfth-century church in Abbey Dore (Herefordshire) has apparently been credited in local folklore as commemorating the place where the wolf corpses were handed over to Edgar.[50] Hundreds of wolf skulls supposedly discovered during the construction of the railway between Brecon and Hereford were the subject of rumours that they were the lost wolves of Edgar and Idwal.[51]

The at least 150-year remove of Malmesbury's account from the event it records calls its veracity into question, especially given that there are no records of sustained attempts to eradicate wolves in England and Wales from the period itself.[52] Although the story does fit with the dearth of wolves in the Welsh archaeological record, wolf remains are rare everywhere – the absence of evidence is not evidence of absence. The only somewhat compelling evidence which may support Malmesbury's account is the fact that places named for wolves during the early medieval period are far scarcer in southern England than in the north, which could suggest that wolves were driven out of the lowlands of the south by the end of the early medieval period.[53] If wolves had been or were in the process of being eradicated from southern England, then it would have been desirable to ensure that they could not recolonise from neighbouring areas. Even so, it seems unlikely that wolf eradication was so fervently desired that their corpses would be demanded over far more valuable forms of tribute. It seems even less likely that as many as 900 wolves existed in Wales to be killed in the first place, nor that the species was eradicated from the entire country in the space of just three years.

It is also often repeated that King Edgar allowed criminals to avoid punishment if they paid for their wrongdoing with wolf tongues and heads. But the source of this information is even later than William of Malmesbury, originating with a remark by English poet Philip Sidney in 1577.[54] Sidney appears to have invented the story based on Malmesbury's depiction of Edgar as a ruler so stringent in his pursuit of law and order that there were almost no thieves or robbers in England. As Malmesbury asks: 'how could a king overlook the criminal acts of men, if he had it in mind to exterminate from his kingdom even those beasts of every kind that shed blood, and laid on Idwal king of

Figure 3.1. Front panel of the St Andrews sarcophagus, depicting an amalgam of various scenes. The hunting scene appears in the bottom left, with the wolf in the bottom leftmost corner, and the dog and hunter behind him. SC 341626 © Crown Copyright: HES

the Welsh an obligation to pay him an annual tribute of three hundred wolves?'[55]

Although they were almost certainly not slaughtered on the scale described by Malmesbury, wolves were hunted from time to time using a variety of methods. One eleventh-century English text features narration from a hunter who describes using dogs to hunt various animals, including wolves, sometimes using the hounds to drive the quarry into nets. A similar method, known as hunting *par force*, involved the pursuit of a single animal by hunters with hounds, and may have been employed in Scotland. A scene carved on the St Andrews sarcophagus – a Pictish monument from the turn of the ninth century – depicts a dog accompanying a hunter, who brandishes a spear at a slavering wolf which appears to be running away (Figure 3.1). This is the only early medieval illustration of wolves being hunted *par force* in Britain, however, and may not be an accurate reflection of wolf-hunting in Scotland at this time.[56]

A similar hunting method involved driving an animal towards waiting bowmen, or else towards pits into which they fell. This tactic was probably the most common method of deer hunting used in medieval Britain, and may also have been used to trap wolves – there are numerous references to 'wolf pits' in English and Welsh documents and place-names from this period.[57] This method of wolf removal could also be used passively which, requiring little skill and experience, allowed communities to reduce local wolf numbers or remove a problem wolf without special expertise.[58] That said, there is little physical evidence to support the existence of wolf pits. Moreover, there are references to 'giant pits' in early medieval English records,[59] which may suggest that 'wolf pits' instead 'allude to one or other of the metaphorical sorts of wolf which haunted the Anglo-Saxon mind', as natural historian Oliver Rackham suggests.[60]

Whatever the methods used to kill them, persecution and habitat loss ensured that wolves were apparently largely absent from southern England by the eleventh century.[61] In the areas where they did survive, wolves were likely forced into more marginal and remote habitats far from settlements,[62] although in parts of northern England wolves still appear to have lived in the same places where sheep were intensively farmed,[63] perhaps leading to clashes with local people.

One fairly reliable retreat for a persecuted wolf was woodland.[64] Deer – especially roe, which was the most ubiquitous deer species in England at this time, as today – likely spent much of their time in or near woodland, which will have attracted wolves to such areas.[65] Though these prey animals were also targeted by people, hunting may have been restricted in some places (particularly estates) to those with especial rights, potentially reducing human footfall.[66]

Likewise, although by at least the seventh century much of the woodland left in England was coppiced and exploited for building material, fuel, and sometimes pasturage, forests were often far from settlements, and woodland pasture appears to have been located primarily on the edges of woodland rather than deep within the trees. Human activity was also restricted throughout much of the year in some woodlands, and larger forests saw less disturbance. This granted wolves some reprieve from the pressures exerted on them by an expanding human population.[67]

In Wales, the comparatively smaller human population lived primarily in the lowlands and along the coast, in areas where woodland had been cleared. Most woodland was concentrated in the uplands further inland, and although these areas were used for livestock grazing in the summer, it seems likely that wolves who retreated to such places would be relatively undisturbed. Likewise, settlements in Scotland were confined largely to the lowlands of the east and southwest, and although woodland had been replaced by bog and moorland in parts of the uplands, there were still patches of forest where wolves could seek shelter.[68] Ireland's woodlands were also exploited for building material, fuel, and food, with one eighth-century law valuing land not only based on whether it could be farmed but also whether it was close to woodland. Ironworking, a particularly ubiquitous activity in early medieval Ireland, was also carried out close to woodlands for easy access to fuel.[69] Precisely because it was such a valuable resource, however, woodland was protected from overexploitation, suggesting that wolves may have found sanctuary within.[70]

The usage of woodland by wolves perhaps led to the common association between these animals and forests in early medieval English literature, a conceptual link which is still prevalent today. Several Old English poems describe the wolf as a creature of the *weald* ('forest') or *wudu* ('wood'), while one explicitly calls it a *holtes gehleða* ('companion of the woods').[71] One poem called *Maxims II*, which presents supposed aphorisms regarding the nature of the world, even expresses that 'the wolf must be in the wood'.[72]

But this sanctuary was slowly shrinking. Woodland coverage had already been reduced significantly by the early medieval period, especially in the south and east of England, and continued to diminish until trees covered only 15% of the country by the time the Normans arrived.[73] Likewise, more than half of Scotland's forests had already been felled many centuries prior, a trend that continued in the early medieval period, while woodland clearances took place in Wales throughout the Roman and early medieval periods and beyond.[74] Although some wooded areas in Ireland were probably maintained to ensure a supply of resources, there is evidence that clearance continued into – and peaked in – the ninth century, at which point records indicate that only three substantial areas of forest still remained.[75] As one of the environments with the greatest 'carrying capacity' for wolves, the loss of woodland would have significantly impacted upon local wolf populations.[76]

Moorland and the generally less heavily settled uplands were likely also important wolf habitats, although both environments were exploited for pasture and transhumance (the movement of livestock to graze in the uplands in summer and the lowlands in winter), hunting, mining, and resource gathering.[77] Textual evidence hints towards at least a conceptual association between moorland and wolves in England at this time: the wolf in one Old English poem known as *The Fortunes of Men* is termed a *har hæðstapa* ('grey heath-treader').[78]

Fenland probably provided another useful refuge for wolves, despite being used for pasture, fishing, fowling, the production of salt, obtaining fuel from peat, and growing crops on the especially fertile reclaimed (drained) land.[79] However, the use of fenland as pasture was seasonal, settlements were far apart, and travel was challenging because of the fluctuating levels of the water.[80] Hunting in marshy areas was also vastly more difficult for people than it was on dry land,[81] giving wolves who lived in wetland some reprieve to enjoy their diet of wading birds, small mammals, frogs, and fish in relative peace.[82] That wolves did inhabit fenland is attested by the numerous place-names in such areas which have a 'wolf' element, as well as their depiction as fenland-dwellers in Old English literature.[83] The wolf-like monsters of *Beowulf* live in swampy marshland surrounded by *wulfhleoþu* ('wolf-slopes'), while the elusive poem known as *Wulf and Eadwacer*, whose titular character (Wulf) bears a lupine name, is also set in the fens.[84]

Wolf habitats and home ranges will not have been fixed, however. Packs moved across different landscapes and travelled nearer or further from human settlements depending upon a range of factors, including prey availability, inter-pack competition and conflict, relationships with humans, and seasonality.[85] In the winter months, reduced wild prey numbers in combination with the removal of livestock nearer to settlements may have drawn wolves closer to human centres to risk an easy meal. But in the spring, when they had pups to protect, they probably stayed well away from people.[86]

Wolves Make Their Mark on the Map

This fragmentary picture of the habitats occupied by wolves in early medieval Britain and Ireland is supplemented by evidence from place-names (also called toponyms), some of which survive today and some of which are no longer used, existing only in historical documents. Many place-names were bestowed based on topographical, faunal, or floral features, reflecting an intimate knowledge of these landscapes as well as the creatures that lived – or were perceived or believed to live – within them.

Few animals have lent their names to as many locations across England as the wolf, but the prodigious number of lupine place-names may not merely suggest that wolves were widespread at this time.[87] That so many places were named for wolves bespeaks a close awareness of this animal's presence in the landscape, perhaps suggesting that knowledge of wolf movements and territories was advantageous, especially if they posed – or were considered to pose – a risk to livestock. It could even signify a simple fascination with these animals.

Around 230 places in England that were named from 450 CE onwards bear a name including 'wolf' or 'whelp' (a term for a wolf pup), and even more are named for people whose names included the lupine elements 'Wulf' or 'Ulf'.[88] In comparison, fox-related place-names number just over 200, red deer place-names number 185, and badger place-names total 141. Even livestock place-names – which, because they relate to animals far more intimately entwined with human lives, are more abundant than places named for wild creatures – barely compete with the lupine toponyms. Cattle only slightly edge wolves out, with 257 places named for them, but wolves drastically outcompete sheep and pigs, which each have only around 125.[89] Ironically, it is easy to mix up wolf and sheep place-names, since some 'wolf' place-names have mutated into 'wool' over the centuries,[90] reflecting a changing landscape which housed more and more sheep and fewer and fewer wolves.

These place-names give valuable insight into the types of landscape with which wolves were associated during the early medieval period, whether culturally or in reality. Wolf place-names were most commonly given to fields, hills, valleys, high ground, clearings, and woodland. Others refer to 'wolf pits', perhaps denoting holes used to trap wolves, as well as 'wolf enclosures' where livestock were housed to protect them from wolf depredations.[91]

Although wolf place-names are encountered throughout most of England (Figures 3.2 and 3.3), the majority are concentrated in the much hillier north and west (in Cumberland, Westmorland, and West Yorkshire), where settlements were usually only found in the pockets of lowland which interspersed the uplands.[92] The generally sparser examples of lupine toponyms from the much flatter landscapes of the

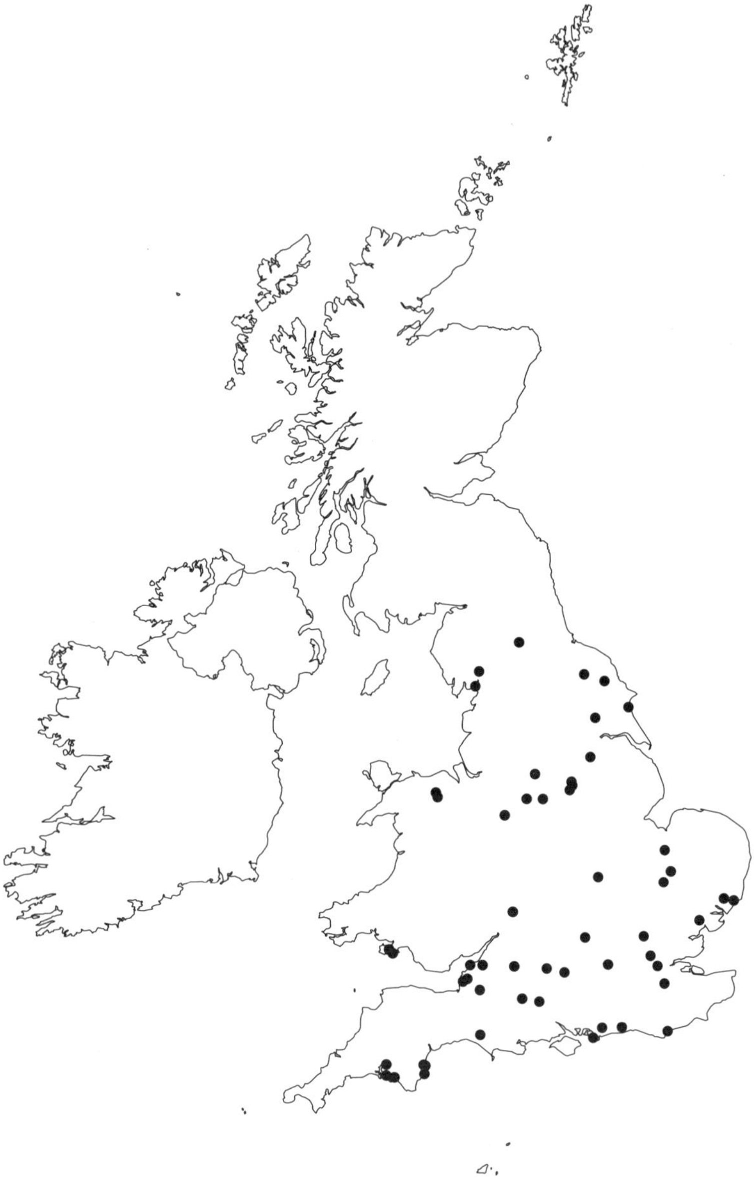

Figure 3.2. Distribution of wolf archaeological finds in Britain (redrawn from C. Aybes and D. W. Yalden, 'Place-name Evidence for the Former Distribution and Status of Wolves and Beavers in Britain', *Mammal Review*, 25 (1995), 201–27, p. 205).

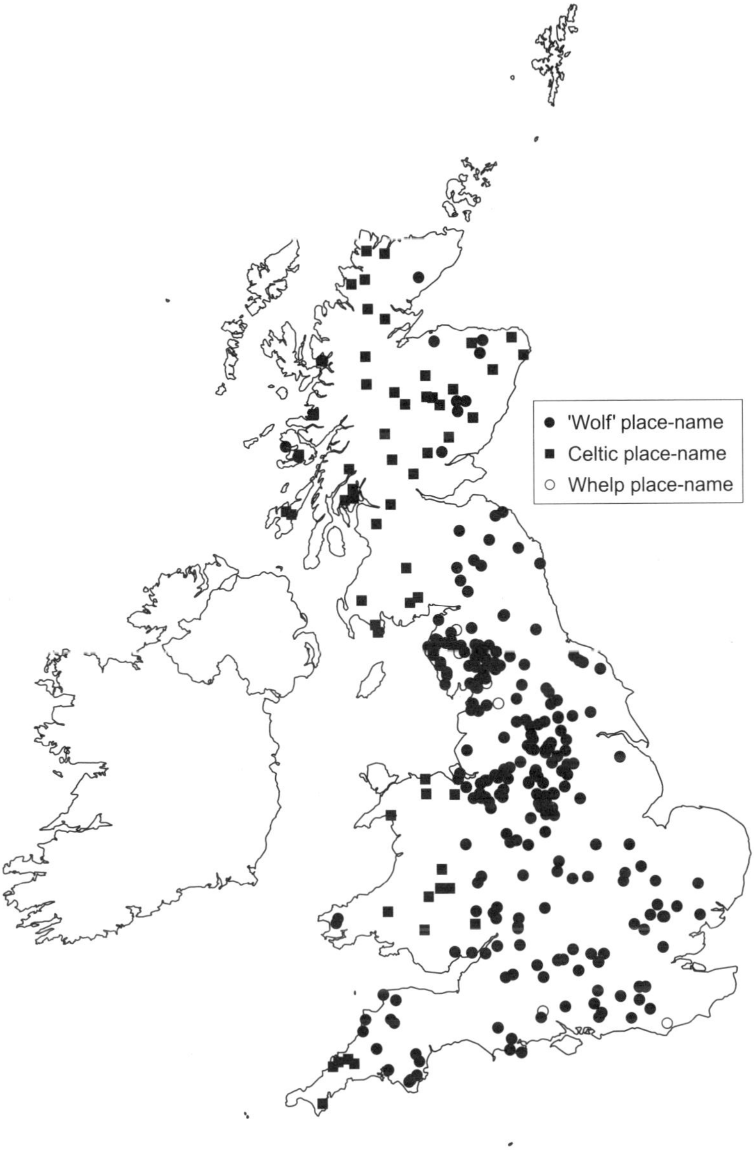

Figure 3.3. Distribution of wolf place-names in Britain (redrawn from Aybes and Yalden, 'Place-name Evidence for the Former Distribution and Status of Wolves and Beavers in Britain', p. 205).

south and southeast, meanwhile, may indicate that wolves had largely disappeared from these areas as early as the sixth or seventh century (the lack of lupine place-names suggests that there were few wolves present to lend their names to the landscape).[93]

Old English *wulf* also appears in a few toponyms in Wales, where there are numerous places also named with the Welsh for 'wolf', *blaidd* or *bleidd* (which also appears in one Herefordshire place-name),[94] as well as *bleiddyn* ('young wolf'), *pothan* ('wolf cub') and *cnud* ('wolf pack').[95] Given the concentration of *blaidd* place-names in the central uplands, it has been suggested that wolves were isolated to these mountainous areas before their disappearance from Wales in the later medieval period.[96] The variant *bleit* also appears in a handful of Cornish place-names due to the area's Brittonic heritage.[97]

Wolf place-names are to be found throughout Ireland (Figure 3.4), the most common of which include the Old Irish (also known as Old Gaelic) terms *bréach* and *mac tíre*, the latter literally meaning 'son of the country'.[98] Wolves were also termed *fáel* and *glaídem* (both meaning 'howler', the latter perhaps related to the Scottish Gaelic *gladaman*),[99] as well as *cú allaidh* ('wild dog').[100] These toponyms are encountered across Ireland, and it is likely that many other place-names which refer to 'canines' more broadly were also related to wolves.[101] Lupine names were frequently given to 'townlands',[102] a word that may lead us to think of human settlements today but which actually refers to units of land as per the land division system in place in Ireland since the early medieval period. Other lupine place-names refer to hills, mountains, valleys, fields, and forts,[103] the latter perhaps because ringforts may originally have been built to keep wolves (as well as other unwanted interlopers) out.[104] There are also a handful of 'wolf' place-names in Ireland, though they are of more recent origin – the product of later English invasions.

Around 100 place-names in Scotland include a 'wolf' element. Scottish Gaelic features several words for 'wolf',[105] the most common of which is *madadah* (which can also be used of dogs, with the distinction sometimes drawn by attaching the word *allaidh*, meaning 'wild'), as is reflected by the 80 or so Scottish lupine toponyms including this element.[106] Other more obscure Gaelic words are *choille(-chù)* ('forest dog'), *cù-fàsaich* ('wilderness dog'), and *gladaman* ('howler'),[107] while *bréach* was shared with early Irish.[108] Lupine names were often given to hills in Scotland, along with corries and hollows, fields, streams,

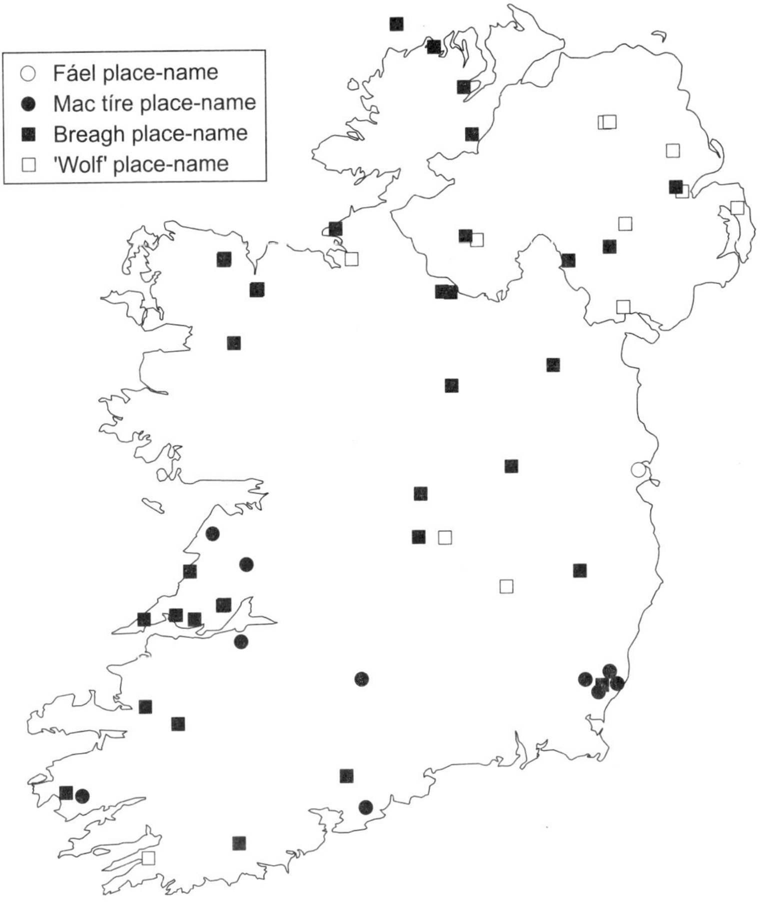

Figure 3.4. The distribution of wolf place-names in Ireland.

and glens. These names are scattered throughout Scotland, although there is a particularly concentrated cluster around the west coast and what is now the Cairngorms National Park, perhaps reflecting the sparseness of human settlements in these upland areas. In the Scottish borders, meanwhile, there are some place-names that also include the English 'wolf', such as Wolfhopelee and Wolflee, both just north of Northumberland National Park.

Pictish stones may also indicate the sorts of places frequented by wolves. Found throughout eastern Scotland, these monuments are carved with symbols and animals. In the absence of written culture they may have functioned as a method of communication, acting as signposts or providing useful depictions of the landscape one was entering into at the edge of territories. A fish might represent a river or body of water, for example, while an eagle could represent mountains. Perhaps a wolf may have been a symbol for woodland, though unfortunately many of the stones have been moved to different locations over the years and, in many cases, their original sites went unrecorded. Another theory is that the wolves on these stones were hieroglyphic symbols representing the names of elite warriors, with the stones themselves being burial or territory markers of individuals or groups who styled themselves as 'wolves'.[109]

One of these stones indicates a particularly close familiarity with wolves. This monument, known as the Ardross wolf (Figure 3.5), is famous for its exceptional realism. Carved during the seventh century and found just north of Inverness, the consensus is that whoever brought this stone wolf to life with chisel and hammer was intimately familiar with the real thing. The creature in question is captured mid-lope, ostensibly about to take a step off the right-hand

Figure 3.5. The Ardross stone, with its famous wolf carving.
© Ewen Weatherspoon on behalf of Inverness Museum and Art Gallery, High Life Highland.

side of the stone. Curved lines may suggest the colour pattern of the fur,[110] and its legs even feature a visible carpus, the joint that connects the foot to the leg bone, lending a particular air of realism to this lupine sketch. This wolf may be stalking a deer – a second stone that appears to depict this prey animal may once have been attached to the first.[111]

'Dark Age' Wolves

Where wolves were present, they could not be ignored. This much is as true today as it was a thousand or more years ago. But how true is the stereotype of the early medieval period as a time when people lived in fear of being eaten alive by slavering, bloodthirsty beasts who prowled the deep, dark forests of their dominions? A Dark Age haunted by dark creatures. The truth is certainly far more nuanced, but difficult to uncover in the midst of so much mythologising.

According to one frequently cited legend, first set down in the seventeenth century by an Anglo-Dutch writer by the name of Richard Verstegen:

> The month which we now call *January* [the Anglo-Saxons] called Wolf-month [...] because people are wont always in that month to be in more danger to be devoured of wolves, than in any season else of the year; for that through the extremity of cold and snow, those ravenous creatures could not find of other beasts sufficient to feed upon.[112]

The Scottish Gaelic name for the same month (*faoilleach*) and the Irish term for the first 15 days of February (*faíleach*) have likewise been speculatively related to Gaelic *fáel* or *faol* ('wolf'), and have been interpreted to mean 'the wolf-time' or 'the month of wolf ravaging'.[113] If so, these names may be attributable to increased human–wolf contact in the colder months, when the scarcity of natural prey may have drawn wolves closer to settlements in search of food,[114] perhaps leading to greater fear of the animals during the winter (though not necessarily a greater threat from them, contrary to Verstegen's claim).[115] Alternatively, these names could be related to the seasonality of wolf hunting, especially given that later medieval hunting manuals do speak of a 'winter season'.[116] Just as likely, however, is that Verstegen's claim is the result of retrospective mythologising, as is so often the case when it comes to wolves.

It is also often said that so-called spitals – hospices that dotted the landscape of early medieval Britain – were constructed as places of refuge where travellers could shelter from the dangerous wild animals that stalked the wilderness. One such spital in North Yorkshire is the subject of a charter penned during the reign of Henry VI (1421–71), which describes that the hospice was built in the tenth century 'for the preservation of the people passing that they should not be devoured by wolves'.[117] As researcher Tim Flight notes, however, this charter was written as supporting evidence in a land ownership dispute, a claim which relied on the supposed 'ancient date' of the spital's construction.[118] Spitals did exist in the early medieval period, but their purpose was not so specific – they were simply places where travellers could take some rest, safe from not only wild animals but from the elements and from a far greater threat: nefarious people.[119] It is also highly unlikely that the etymology of the word 'loophole' derives from holes through which people staying in spitals could spy for wolves (*loup* in French).[120] The true origins are rather less sensational, lying in the medieval Dutch verb *lûpen* ('to lie in wait'), and referring to wall slits through which arrows could be shot.[121] The myth of the wolf spital is an invention based on conventional ideas of the 'Dark Ages', an imagined primitive world in which wolves terrorised weary pilgrims traversing desolate wildernesses and drove them to seek shelter from their snarling jaws.[122]

The reality was far less frightening than such horror stories would suggest. On a daily basis it is likely that wolves were simply afforded the same 'healthy, health-preserving respect' as any other wild and potentially dangerous animal.[123] Although wolves were probably somewhat less wary of humans than today (since firearms, which have contributed to the wolf's fearfulness of humans, had not yet been invented),[124] they are unlikely to have been responsible for anywhere near the numbers of attacks which they are often assumed to have perpetrated. There is little contemporary evidence of attacks which, if they did occur, were probably isolated, opportunistic incidents rather than common occurrences.[125] Humans certainly did not form a significant part of the wolf's hunting repertoire nor diet. Rabies, moreover, which is the cause of most wolf attacks both today and probably in the past,[126] does not appear to have been a significant issue in early medieval Britain and Ireland.[127] Most meetings between wolves and humans likely saw the animals fleeing or, if the wolf was not lucky enough to escape, far more damage and pain inflicted upon them by humans than the reverse.

If people did fear wolves, it is likely that such fears were based in cultural exaggeration rather than real-world experiences. One such exaggeration is a trope commonly seen in Old English, Brittonic, and Scandinavian poetry known as the 'Beast of Battle' topos (a traditional formula seen throughout a corpus of literature). This topos depicts wolves, eagles, and ravens eating corpses in the wake of battles, or else as heralds of doom who appear before the battle begins, in anticipation of the impending slaughter and subsequent feast. In a scene depicting a nation's downfall and devastation at the end of *Beowulf*, for instance, it is said that 'the dark raven' 'plundered the slain with the wolf',[128] while in a poetic retelling of the Old Testament Book of Exodus, wolves are seen singing a 'horrific evening-song in expectation of food'.[129] Similarly, one Brittonic poem called *Y Gododdin* – a collection of elegies celebrating the fallen warriors of the northern kingdom of Gododdin, written around the sixth or seventh century by a court poet named Aneirin – describes a battle as a 'wolf-feast', while one warrior is described as having 'fed the wolves by his hand' (that is, provisioning a feast with all the corpses he cut down).[130] Wolves are also depicted eating human bodies in several Irish texts. In one particularly gory example from the eighth-century tale of *Togail Bruidne da Derga* ('The Destruction of Da Derga's Hostel'), a warrior is eaten alive as he lies on the battlefield, the wolf's entire head buried within the wound.

Wolves are also frequently depicted eating humans even outside battle. In *Maxims I*, another Old English poem of collected aphorisms, the wolf is described as 'wailing for hunger', an animal which 'certainly does not mourn the slain or the slaughter of men, but he always wants more'. *The Fortunes of Men* depicts being eaten by a wolf as one of the fates which can befall a young man as he sets out on his journey to adulthood. In an Old English homily recorded in the tenth-century Vercelli Book manuscript, a person's soul addresses their body and denigratingly refers to it as the 'food of worms and rending of wolves and tearing of birds'.[131]

Although such images are probably based in truth,[132] the regularity with which wolves actually engaged in this activity may have been far less common than the poetic record would make it seem. It is also only partly reflective of the reality that many animals likely joined in the post-fight feast, including crows, foxes, other birds of prey, and even cats and dogs.[133] In any case, the opportunistic consumption of human flesh would not have given wolves a taste for human meat which led them to actively hunt people, since wolves do not perceive live prey in the same way as they do corpses.[134] The absence of any records of wolf attacks on people, despite the wealth of annals chronicling events of the period, suggests that attacks were incredibly rare.[135] As today, people in medieval Britain and Ireland had a greater taste for lupine horror stories than wolves did for human flesh. Wolves were 'a greater threat to the medieval psyche' than to the medieval person.[136]

Wolves Old and New

The nature of this psychological threat was directly informed by the seismic cultural and religious changes that occurred during this period. After the Romans departed, the Britons continued to practice Christianity. The Germanic settlers, Picts, and Gaels were all 'pagan',[137] on the other hand, though they had different – and, in the case of the Picts and Gaels, less well-understood – spiritual beliefs. These paradigm shifts shaped people's thoughts and daily lives and, in turn, affected the treatment of wolves in both legend and landscape.

For one sixth-century Briton, a monk named Gildas, the arrival of the pagan Germanic invaders felt like the biblical metaphor of the wolfish devil attacking Christian sheep had come to life. In his *On the Ruin and Conquest of Britain*, Gildas laments the appearance of this 'ferocious' people 'hated by man and God', who arrived 'like wolves into the fold' and proceeded to lay waste to everything in their path.[138]

For the Germanic settlers, however, wolves meant something very different. From their perspective, to be wolfish was to be not depraved and devilish, but strong and courageous. Their warriors might even deliberately emulate the wolf's characteristics, wearing wolf pelts to 'transform' into the animal during battle or to embody its power by channelling an 'inner wolf'. Such warriors, known as *ulfheðnar*, were often associated with Odin, a Norse god linked to shapeshifting, fury, war, and death. Odin was frequently accompanied by two lupine companions named Geri and Freki, and was ultimately eaten by the monstrous wolf named Fenrir during Ragnarǫk, the apocalyptic cataclysm foretold in Norse mythology during which the world would be destroyed.

Geri and Freki are themselves depicted as 'Beasts of Battle' in one Old Norse poem, prowling the battlefield in the hopes of obtaining flesh to eat – a behaviour certainly appropriate for animals whose names both translate as 'greedy'. Odin himself 'fed' these wolves in a sense, by providing them with a glut of corpses in his capacity as god of death and war – a symbiotic relationship in which the god received the souls of the dead and the wolves got their corpses.[139] In this way, to end up in the stomach of a wolf was no bad thing. Rather, it was an indication that the person had died the glorious death of a warrior and would soon be welcomed to Valhalla by Odin and his wolves. Odin needed the very best warriors to help him defend Valhalla against Fenrir, who patiently 'watche[d] the abodes of the gods', waiting to attack.[140] In this worldview, the ravenousness of wolves and their association with death was simply part of the natural and cosmic order.

Odin was known as Woden in pre-conversion early medieval England. Although there may have been significant differences between the god worshipped in sixth-century England and the iteration recorded in the later Scandinavian sources from which we gain most of our knowledge of him (many of which were put to parchment by the twelfth- to thirteenth-century Icelandic historian Snorri Sturluson), it seems that Woden was also associated with wolves. Wansdyke and the nearby 'Woden's Barrow' in southwest England were perhaps named for this god because wolves and ravens feasted on the corpses left behind in the aftermath of battles fought in the area in the sixth and eighth centuries.[141] Parallels between artistic representations of Woden and Odin on artefacts found in England and Scandinavia also suggest that the two versions of the god – and, in particular, his relationship with wolves and wolf-warriors – were not symbolically dissimilar. A seventh-century mould used for embellishing metal found in Cambridgeshire depicts a spear-wielding warrior with the head of a wolf. This image is remarkably similar to that seen on a sixth- or seventh-century mould from Sweden, implying that Woden, like Odin, was associated with wolf-warriors. Perhaps such warriors were even seen stalking the battlefields in England, garbed in their wolf pelts and howling in their battle-frenzy.[142]

The best-known potential representation of Woden is found on a purse lid unearthed at the famous sixth- to seventh-century Sutton Hoo ship burial in Suffolk (Figure 3.6). Sutton Hoo is believed to have served as an inhumation site for the Wuffings, a dynasty who ruled East Anglia during the early medieval period.[143] The name 'Wuffing', a variant form of 'Wulfing', means 'kin of the wolf', and the dynasty may have held a 'totemic affinity' with this animal.[144] The metallic Sutton Hoo ornament features a man sandwiched between two stylised wolves who seem to be whispering into his ears, perhaps representing Geri and Freki speaking into Woden's ears in the same manner as the god's ravens, Huginn and Muninn, were said to bring him wisdom and secrets from their travels.[145] It is also possible that these lupine figures represent wolf-warriors,[146] worshippers of Woden/Odin who hoped to fight alongside him at Ragnarǫk against the great wolf, Fenrir.[147] Alternatively, the wolves could represent a totemic lupine

Figure 3.6. Detail from the Sutton Hoo purse lid, depicting a man between two lupine figures.

Figure 3.7. Medieval East Anglian coin depicting a wolf, in the same style as the Icenian coins. Spink Auction 21000, 18 March 2021, Lot 71.

protector – an 'ancestral guardian-spirit' who flanks the warrior to keep him from harm.[148]

Coins produced in East Anglia during the early eighth century also depict wolves (Figure 3.7), perhaps due to the emblematic significance of this animal in the region. Visual similarities indicate that these coins were copied from the same lupine Iron Age coins found in Norfolk which we saw in the previous chapter.[149] Perhaps an early medieval East Anglian unearthed one such coin, delighted to discover that it bore the very same animal which held such importance to their people.

Although perhaps best known for its representation on the Franks Casket, an eighth-century box made from whalebone found in Northumbria, the motif of the she-wolf suckling Romulus and Remus was popular in East Anglia, perhaps because, as academic Sam Newton puts it, 'the essentially totemic Roman foundation-legend of the she-wolf and twins appeared congruent with East Anglia's own, arguably totemic, ancestral associations with the wolf'.[150] Romulus, Remus and the she-wolf are seen on coins issued under the late eighth-century East Anglian king Æthelberht II, whose design is probably copied from Roman coins which circulated in England during the fourth century CE.[151] The wolf and twins also appear on an eighth-century whalebone panel found in Norfolk which may have adorned a book, writing tablet, or shrine,[152] as well as a gold bracteate (a flattened piece of round metal with a stamped design, worn as a necklace or amulet) dating to around 475 CE, which was uncovered in the fenlands of Suffolk.[153]

Christianity re-emerged as the dominant belief system across Britain and Ireland following successful missions from mainland

Europe between the fifth and seventh centuries. But even then, the Wuffings did not abandon their lupine connections despite the wolf's negative symbolism in Christian theology. Pagan and Christian associations with the wolf are both seen in a tenth-century story recording the death of King Edmund of East Anglia (r. *c*.855–69), who seems to have been the last king of the Wuffing dynasty.[154] In this tale, East Anglia is invaded by a devilish group of pagan Vikings who are likened to ravenous wolves. They decapitate the pious Edmund when he refuses to renounce Christ, dumping his head in the woods. Miraculously, the head comes back to life and reveals its location with its cries of 'Here! Here! Here!'. Following his voice, Edmund's people find the head being guarded by a tame wolf who clasps it between its paws, its hunger held at bay by the power of God even as blood oozes from the severed neck. Though the description of the Vikings as wolf-like taps into the New Testament comparison of the devil to the wolf, it is likely that the story is partly owed to East Anglian traditions of the animal as a totemic guardian,[155] especially since the wolves on either side of the man on the Sutton Hoo purse lid are reminiscent of the way in which Edmund's lupine protector grasps his head between its paws.[156]

The metaphor of the devil and devilish people as wolves that is seen in this tale was ubiquitous throughout Britain and Ireland. One of the most famous artistic representations of the wolf from this period, found in an early ninth-century Irish manuscript known as the Book of Kells, sports a fiendish forked tail – a visual representation of the devilish wolf of the Gospels which it appears alongside (Figure 3.8).

The wolf metaphor was especially popular with clerics, who employed it to emphasise the godlessness and perceived savagery of those who ignored Christ's teachings. One tenth- to eleventh-century English monk named Ælfric describes how 'dishonourable criminals and deceitful thieves' are punished by society and God alike, 'because they lived by plundering like savage wolves, and they often took from the righteous their sustenance'.[157] An Old English poem about St Andrew, meanwhile, takes the metaphorical depredation of the wolfish devil on the 'sheep' of Christ's flock and makes it gorily and cannibalistically literal, depicting a race of heathen 'slaughter wolves' who drink the blood and devour the flesh of their sheep-like victims, whom they drug and cage like livestock in a pen. Patrick, the fifth- to sixth-century bishop and patron saint of Ireland, similarly likened

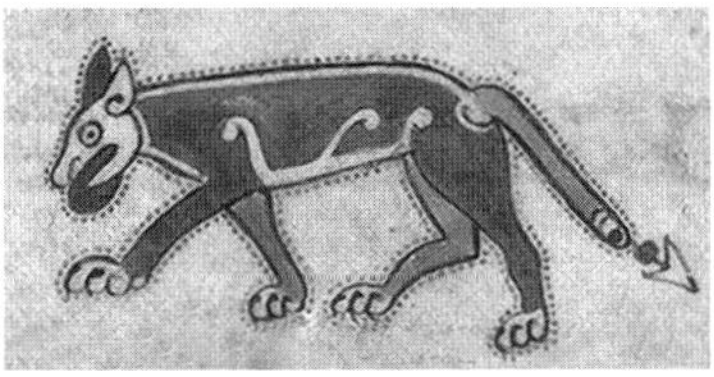

Figure 3.8. Illumination accompanying the Gospels in the Book of Kells, depicting a wolf with a forked tail. By kind permission of the Board of Trinity College Dublin.

'father-slayers' and 'brother-slayers' to 'savage wolves devouring the people of God as they would bread'.[158]

But wolves were not always imagined as slavering, devilish beasts in early British and Irish literature, perhaps a product of inherited Celtic spiritual beliefs which treated wolves more favourably than the Christian tradition. Irish, Scottish, and Welsh saints are frequently seen taming wolves, overcoming or suspending the wolf's ravenousness in a representation of the power of God and faith within Him. These saints often compel wolves to replace the animals they have eaten, or else to go against their predatory nature by guarding livestock. A wolf who ate a calf in the presence of sixth-century Irish saint Finan, for example, brought a magical white calf as a replacement before committing itself to serve as a dog-like guardian of the herd. St Patrick's prayers supposedly caused a wolf that had stolen a sheep to return its loot – right from within the wolf's maw according to some accounts. On other occasions, the wolf takes the place of the animal it has eaten. Cainnech, an Irish saint who preached in Scotland and Ireland in the sixth century, was said to have compelled a wolf to fill in for a calf it had consumed, allowing its mother to continue to produce milk. Similarly, when the sixth-century patron saint of Glasgow, Mungo (also known as Kentigern), lost one of his tamed ploughing deer to a wolf, his prayers prompted the wolf to return and take the stag's place. Though such stories suggest that wolves were considered a nuisance for their tendency to steal livestock, they also reflect more empathetic responses to depredation. Even in secular tales, wolves are not subject to lethal methods of control. In the *Togail Bruidne da Derga*, the legendary king of Ireland Conaire Mór keeps seven wolves hostage, to ensure that their wild counterparts do not prey on too many cattle.

Some saints are even depicted showing compassion towards or befriending wolves. The sixth-century Irish saint Molua was said to have established an annual tradition in which a pack of wolves were gifted a calf by the brothers of a monastery, in exchange for which the wolves protected the monastery's sheep from both thieves and other wolves. Another early medieval Irish saint, Máedóc of Ferns, sacrificed eight sheep to feed eight hungry wolves, for which he was rewarded by the miraculous appearance of eight replacement sheep. Sixth-century Irish saint Ailbe saved a female wolf from hunters before providing food for her and her pups. Ailbe later employed a pack of wolves to protect his monastery against thieves, in return for which he provided the wolves with shelter, washed their feet, and gave them a calf to eat, a feast that took place every year. Ailbe had a particular affinity with wolves, having been suckled by one after being abandoned in the wilderness as an infant, an act of love which left the wolf grief-stricken after Ailbe was rescued. The legendary Irish king Cormac mac Airt was also said to have been suckled by a female wolf as an infant, according to one eighth-century tale.[159] Wolves were not just shown kindness by saints – they were capable of acts of kindness themselves.

Wolfish characteristics were even sometimes admired in the Old North. Brittonic poets often celebrated the perceived wolfish qualities that lent warriors success in battle, such as strength, cunning, ferocity, and bravery. One sixth-century Welsh-Brittonic poet called Taliesin frequently makes use of the wolf-warrior metaphor even despite apparently being Christian, such as when he celebrates a warrior named Owain who 'punish[ed]' his opponents as 'severely [...] as wolves punish sheep'.[160] In *Y Gododdin*, Aneirin praises the warriors of Gododdin whom he describes variously as a 'wolf in fury', a 'wolf of the host', and as a 'man who used to slaughter in wolfish rage'.[161] One warrior is even lauded for favouring lupine violence over the civilised customs of humans, being 'quicker to a wolf-feast than to a nuptial'[162] – that is, more eager to attend a battle than a wedding. These warriors are also praised for 'feeding' the wolves by cutting down their opponents. The greater the warrior, the fatter the wolves became.

The acceptability – or even desirability – of wolfish characteristics in certain contexts may be reflected in medieval naming practices. People – usually but not exclusively men – were often given names that included a 'wolf' element, such as the English names Sigewulf ('victory wolf'), Ecgwulf ('sword wolf'), and Rædwulf ('advice/intelligence wolf'); Welsh Bledig ('wolf-like'); and Irish Cunagussos ('he who has the strength of the wolf'), Conal (derived from the Old Celtic word *cunovalos*, meaning 'strong as a wolf'), and Cenn Fáelad ('wolf's head' or 'head of wolves', implying strong leadership and encapsulating the intelligence and cunning of the wolf). Admiration for the wolf was even more explicitly expressed in the Irish name Conchobhar, an early form of the modern name Connor, which meant 'lover of wolves'. Even following conversion to Christianity such names continued to be bestowed, perhaps in the hope that the recipient would grow up to become a great warrior imbued with the attributes of his namesake animal.

Clan names and surnames in Scotland also sometimes featured the wolf. The Gaelic kingdom of Dál Riata in western Scotland was ruled at one stage of the early medieval period by the Cenél Loairn, a kin group thought to have originated with the eponymous king Loarn mac Eirc, whose first name came from a Gaelic word for 'wolf'.[163] The earliest written record of Gaelic found in Scotland, a tenth-century manuscript known as the Book of Deer, also mentions a man named 'Duncan mac Síthech'. The term Síthech, which ultimately morphed into the modern name 'Shaw', means 'wolf'. The suffixes of two other names, MacLellan and Gilfinnan, are thought to derive from the name Fhaolain (*faolan*), another term for wolf.[164]

Some names, such as the English Æthelwulf ('noble wolf') and Eadwulf ('prosperous/blessed wolf');[165] Welsh Bleiddri ('king of the wolves' or 'wolf-like king');[166] and Irish Cunamgli, Cun[a]netas and Cvnorix ('princely/kingly/champion-ly like the wolf') were given to members of the nobility,[167] perhaps to inspire respect, fear, and admiration, and to indicate that the person who bore such a name was a powerful individual, a 'predator' at the top of the hierarchy who ruled over an underclass of sheep-like commoners.[168] In other cases, wolf names may have served a protective function, perhaps by invocating a totemic guardian spirit, such as the English names Wulfhelm ('wolf helmet'), Wulffrith ('wolf peace'), and Wulfweard ('wolf guard/guardianship').[169] Even important figures in the Church could have lupine names, such as the eleventh-century Archbishop of York, Wulfstan ('wolf stone'). Wulfstan did not shy away from the lupine meaning of his name, titling one of his homilies the *Sermo Lupi ad Anglos* ('Sermon of the Wolf to the English'). Yet, significantly, while *wulf* was a popular part of compound names in early medieval England, it was highly uncommon to come across a person named simply 'Wulf',[170] suggesting that the wolf element needed 'diluting' with another term.

Wolfish behaviour was indeed vehemently deplored in some contexts. Criminal outlaws were frequently compared to or described as 'wolves', and deemed to be wolf-like for their depredations upon others (as in Ælfric's aforementioned description of the thieves who 'lived by plundering'). In turn, criminals were often cast out of the community, forced to flee to the same wilderness haunts inhabited by (or thought to be inhabited by) the wolf, further reinforcing perceptions of such people as wolf-like. If a person's legal status was revoked, so too was their status as human.

In Ireland, an undesirable outsider and a despised criminal could be declared a *cú glas* (literally '"grey dog"', perhaps meaning 'wolf'),[171] losing their legal, social, and human statuses. There are also references in Old Irish texts to entire bands of wolfish outcasts. Known as *fían* or *díberga*, these groups of men had left society and civilisation behind to go '(were-)wolfing',[172] roaming the wilderness like packs of wolves, wearing wolfskins as they murdered, plundered, and robbed, hunting animal and human prey alike.[173] In England, meanwhile, in a law purportedly in force during the reign of Edward the Confessor (r.1042–66),[174] an outlaw was said to 'bear a wolf's head'.[175] The gallows is also described in one Old English poem as a *wulfheafedtreo* ('wolf-head-tree') – that is, a 'tree' for criminals to be hung upon. *Maxims I*, in one of its aphorisms, also states that a 'friendless' and 'miserable' criminal outcast is forced to take wolves as his companions, though it predicts that he will not survive long among these treacherous lupine friends, who 'very often' will 'tear him apart' since, as we saw earlier, this animal 'does not mourn the slain or the slaughter of men, but he always wants more' flesh to satiate his boundless appetite.

One's status as 'human' was dependent upon adhering to certain rules of law and standards of behaviour. To act like a wolf was to conceptually become one. The 'transformation' of the criminal into a wolf suggests that 'the notion that we can be *like* wolves in our [...] violence, and appetite' had to be 'forcefully expelled'.[176] The urge to remove the 'wolf in sheep's clothing' from the human community was as natural as the impulse to drive a wolf from among a real flock.

Old Wolves Return

But no matter how many times they were expelled the wolves always returned, one way or another.

In the late eighth century, Scandinavian invaders arrived in Britain and Ireland, raiding and settling until they controlled much of the territory. In England all but one kingdom (the southern kingdom of Wessex) was claimed by the Scandinavians, while in Scotland they controlled Shetland and Orkney as well as the Hebrides and, at times, much of the mainland. Attacks were launched from Scotland into Ireland, where the Scandinavians came to control pockets of land mainly on the east coast (in Antrim, Down, Dublin, Wexford, and Waterford), although they also held territories on the Cork and Limerick coasts, from which they launched raids inland. To the east, the Isle of Man was conquered and settled by the early ninth century, and Wales also saw some raiding from at least the ninth century onwards, although there is little evidence of settlement. While it is generally accepted that the arrival of the Scandinavians was not as drastic nor bloody as popular history suggests, peaceful settlement was accompanied by raids and battles for supremacy.

For those who fought against the invaders, the Scandinavian warriors may have seemed like the lupine Beasts of Battle incarnate, come not only to feast but to prepare the meal themselves.[177] These 'slaughter-wolves', as they are described in a poem which recounts the events of the Battle of Maldon of 991 CE, are tellingly accompanied by circling ravens and eagles who are 'eager for flesh'.[178] The wolves that would otherwise complete the trio are absent, their role instead fulfilled by the human 'wolves' of the Scandinavian army.

The Scandinavian settlers brought old wolves to Christian shores. A wooden amulet found in Dublin, which is carved into an intricate, highly detailed curved wolf who clutches a sphere, appears to represent Fenrir devouring the sun during Ragnarǫk (Figure 3.9). A mid-tenth-century monument found on the Isle of Man, known as Thorwald's Cross, shows Odin being devoured by Fenrir during this apocalyptic battle, while another tenth-century cross from Newcastle upon Tyne depicts Odin's son, Víðarr, avenging his father's death by ripping the wolf's mouth open and killing him. Two stone crosses found in Angus and Perth also feature a man pulling apart the jaws of a wolf-like creature, perhaps another depiction of Víðarr's rending of Fenrir.

Figure 3.9. Amulet found in Dublin, possibly depicting the lupine Fenrir swallowing the sun during Ragnarǫk. © National Museum of Ireland

The St Andrews sarcophagus (upon which we saw the Pictish representation of wolf hunting earlier – see Figure 3.1) likewise depicts the wolves of Ragnarǫk. Here, a mounted Odin grapples with a fearsome-looking wolf (top middle), while to the right a gigantic Víðarr prises apart the wolf's upper and lower jaws with gargantuan hands.

Other sculptural evidence shows that episodes from Old Norse myths were familiar enough in early medieval England that they could be inferred simply from a few choice motifs.[179] A frieze from the Winchester Old Minster (the early medieval church which was replaced by Winchester Cathedral after the Norman invasion) depicts a scene from the life of Sigmund, a member of the legendary Vǫlsung royal family said to be descended from Odin. One episode from this story sees Sigmund and his nine brothers betrayed by the Geatish king Siggeir, the husband of their sister, Signý. Despite Signý's efforts to save them, her brothers are confined in stocks and picked off one by one by the king's mother in the form of a she-wolf. When only Sigmund is left, Signý devises a last-ditch attempt to save him by smearing honey on his face. When the she-wolf comes to claim her final victim, she begins to lick the sweet treat off Sigmund's face, an appetiser to preface her meaty main course. Unfortunately for her, the entrée never comes. Sigmund seizes his opportunity and bites off her tongue, killing her.

Dated to the eleventh century, the Winchester frieze depicts the she-wolf just inches from Sigmund, who is pinned underneath and staring up at her, presumably about to deliver the fatal bite.

But the return of the pagan wolf was short-lived, with conversion eventually taking back all of the pagan pockets of Anglo-Scandinavian England. The Scandinavian stories of these mythological wolves became records of times gone by, of pagan beliefs and stories which, though perhaps enjoyed as fiction and kept alive as a celebration of heritage, had rightfully been replaced by the true, Christian faith.[180] The time of Woden and the wolf-warriors was over.

CHAPTER 4

From Wolf Land to Wool Land

It is dawn, and the dew on the grass is just beginning to glimmer in the low, pale-yellow light of the brisk autumn morning. The king is preparing to set out on a hunt, saddled up on his pristine white horse, accompanied by his most loyal sycophants. They are followed on foot by sleep-deprived servants, some attempting to corral the king's boisterous pack of hunting dogs.

It promises to be a good hunt. A herd of red deer are rutting in the area. The roaring of the stags and the clashes of their magnificent antlers have been heard echoing across the treetops over the past few days.

But your focus is elsewhere. Melding into the trees at the edge of the forest in his brown woollen cloak, a man watches the clamorous hunting party from afar. He too is commencing a hunt, but his is decidedly less grand, and much more dangerous. He is a royal employee. His task: to protect the king's property, the deer that live within this forest.

He is a wolf hunter.

He turns noiselessly, expressionlessly, entering the forest just as the silence is broken by the bellows of horns from the spectacle he has left behind.

His job today is checking his traps. You follow him as he picks his careful but deliberate way through the trees. As the branches envelop you, you become as disorientated as the man is certain of his path.

Deeper into the forest, he finds what he is looking for. A simple contraption: a pivoting beam with a snare at one end to hoist its victim into the air. This one has been tripped, but it is empty. The man deftly resets it before going in search of the second, wondering – as he always does when he finds an empty trap – if his goal is finally complete, the wolfen scourge finally driven from this forest, once and for all.

Navigating his way through the trees once more, the hunter soon locates his second trap.

He smirks.

This time, the rope does not swing empty in the breeze. Something is in it, something distinctly russet-grey, but it is not yet dead.

You hear a pitiful whine that pulls at your heart. You turn away. You wish you could stop this man, stop all the other men in all the other forests who will be the death of this species.

But there is nothing you can do.

William and the Wolves

By the year 1066, a new player in the fate of Britain and Ireland's wolves was on his way. His name was William, Duke of Normandy, but he was soon to be known as William the Conqueror.

William arrived in England to lay claim to the throne in the autumn of 1066. He disputed the succession of Harold Godwinson, brother-in-law to the late king Edward the Confessor (r.1042–66), who had been crowned King of England earlier that year. William met Harold in battle at Hastings on the 14th of October. Harold was killed and, so the stories go, the remains of the English army were left to be picked over by lupine and avian scavengers. Though he was not well accepted by his new subjects, William managed to quash any stirrings of rebellion and was crowned King of England on Christmas Day that same year.

If there was one thing William loved, it was hunting. Borrowing from the Franks, who had established hunting grounds on the Continent as early as the seventh century, the Normans began creating formalised areas for the hunting of deer a few hundred years later.[1] By William's time, this expression of power and status was a much-enjoyed pastime of the elite.

Soon after William became king he enforced this system in England, commandeering huge swathes of land for royal Forests,[2] private hunting reserves where his exclusive right to hunt was enshrined in Forest Law.[3] Vast areas were suddenly taken under royal control, sometimes even including entire settlements. Though they were not seized as royal property, these areas became subject to unprecedented restrictions designed to protect both the deer and their forage (known as vert), which included limiting grazing and tree-felling, and banning the trapping and hunting of all species. Large dogs kept in or near Forests were even required to have their toes removed, to stop them from chasing deer.[4] Within 34 years of William's victory at Hastings, as much as one-third of England's land area may have been claimed as royal Forest.[5]

Along with Forests, the Normans also introduced fenced deer parks, where traps known as 'salters' or 'leaps' – interior ditches set along the fence line – allowed deer to enter but not exit, ensuring a constant supply of quarry for the owner.[6] Both royalty and nobility could enclose land for deer parks, with around 40 such hunting grounds created within 21 years of the Conquest. By the year 1300 a staggering 3,200 deer parks had been demarcated, stocked primarily with fallow deer imported from Europe, which were better suited to these enclosed hunting grounds than the native red and roe.[7]

William protected his new hunting grounds fiercely. A chronicle written after his death describes him as 'fallen in avarice and enraptured by greed', having 'established a great preserve for wild animals and laid down laws for it, that whosoever slew hart or hind was to be blinded'. Though 'his rich men bemoaned it, and the poor men complained' William could not have cared less, 'so unmoved [...] that he recked naught of the enmity of them all'.[8] Nothing could get in William's way – not the disdain of every person in England and, least of all, wolves. He wanted to tame and control the wild,[9] have dominion over all the land and keep its deer all for himself. Wolves had no place in William's world.[10]

But a single country proved insufficient for William's ambitions, and he soon set his sights westward. Although he made little headway during his lifetime, William's successors took up his colonial aspirations, claiming significant portions of land in the Welsh Marches and south Wales by the first half of the twelfth century.[11] They soon parcelled up the land for royal Forests, just as they had in England. Over 100 Welsh Forests were created by the fourteenth century. The largest was the 'Great Forest' of Brecknock (Brecon), where more than 50 square miles were set aside for the hunting pleasure of the nobility.[12]

Although Scotland escaped Norman expansionist ambitions (a peace treaty signed in 1072 ensured that the Normans could not claim any land north of the border as their own), it did not remain untouched by the culture of their new southern neighbours. Prince David of the Cumbrians – who later become King David I of Scotland (r.1124–53) – spent his formative years in the royal household of Henry I of England (r.1100–35), where he was 'saturated by Norman-French culture and social mores', and later gained first-hand experience of the forest system when he became Earl of Huntingdon in 1113.[13] Evidently, he liked the idea. By 1136 the first royal Forests had appeared in Scotland,[14] expanding to over 80 Norman-style Forests and 12 deer parks by the thirteenth century.[15]

Ireland was not targeted by William's descendants until the twelfth century, when the Normans invaded and conquered huge swathes of land from coast to coast, claiming this territory as a Lordship belonging to England in 1177 and leaving only around one third of the island controlled by the native Gaelic Irish.[16] While both Gaelic and Anglo-Norman noblemen enjoyed hunting, only the Normans hunted fallow deer and maintained deer parks. The native Gaels valued open wild landscapes which they could roam over at will, chasing the native red deer.[17] It is unknown how many deer parks and Forests existed in medieval Ireland due to the loss of documents which preserved this information, though surviving records include references to Forests in Limerick, Tipperary, Clare, Westmeath, Wicklow, and Dublin, while non-royal Forests and chases were found more widely throughout the island.[18]

Artificially maintained stocks of deer in the Forests and parks across Britain and Ireland provided a 'tempting larder' that would have been difficult for wolves to resist.[19] But wolves ate these deer on pain of death. It was an insult, a violation of the nobility's sovereignty, theft of what was rightfully theirs. As a result, wolves became the victims of what zoologist Derek Yalden termed 'state-sponsored pest control'.[20] This was not the noble art of hunting by the elite or sporadic efforts to control local populations through trapping, but a ruthless, organised campaign of extermination endorsed by the Norman kings, most notably Edward I (r.1272–1307). This was the beginning of the end for wolves in England and Wales.

Those who hunted and trapped wolves had a variety of methods and tools at their disposal, including nets, pits, traps, and snares.[21] Dogs trained to chase and kill their quarry appear to have been one of the most useful assets to a hunter. These animals are depicted in the Bayeux Tapestry (the 70-metre-long embroidery which records and

Figure 4.1. Detail of a wolf-hunting scene from the eleventh-century Bayeux Tapestry. City of Bayeux.

commemorates William the Conqueror's victory at Hastings), with one decorative scene in the tapestry's margin showing two armed men and four dogs chasing a large wolf (Figure 4.1).[22]

The Norman kings themselves kept hounds for pursuing wolves, employing hunters to make use of them and keepers to care for them. A set of accounts documenting the expenses of Henry I in the year 1136 record wages paid to 'huntsmen of the wolf hunt', for whom horses, dogs, and clothing were provided. The hounds were subdivided into three 'packs', one of which, comprising eight greyhounds and twenty-four racing dogs, was dedicated to 'the wolf hunt'. This pack was entirely separate from the hounds used by the king for his hunting expeditions,[23] suggesting that wolves were hunted not for enjoyment but for extermination and, presumably, that wolves were numerous enough to warrant spending money on dogs and servants dedicating to dispatching them.

The task of maintaining packs of wolf-hunting hounds was also farmed out to the nobility as one of the conditions upon which land was granted to them. Edward I granted land in Essex to a man named William de Reynes in exchange for keeping five wolf-dogs.[24] Edward's son, Edward II (r.1307–27), gave one William Michell land 'by the serjeanty of keeping [the king's] wolf-dogs', and paid another Michell named Richard 4½ pence 'daily from the King's purse ' for 'keeping two wolf-dogs'.[25] Other kings lent their own dogs to the aristocracy so that they could rid their land of wolves. A man named William de Limeres was given land in Southampton by King John (r.1199–1216) in exchange for 'hunting the wolf with the King's dogs'.[26] Landowners themselves could also require their tenants to keep wolf-hunting dogs if they so wished. A twelfth-century landowner in Yorkshire granted pastureland to the monks at Jervaulx Abbey with the express command that 'they [were] to have mastiffs to keep in check the wolves in their pastures'.[27]

The Crown also employed specialist wolf hunters whom they paid handsomely, indicating that a high value was placed on their trade and expertise. Henry II (r.1154–89) gave the sheriff of Hampshire alone a staggering 100 shillings to pay his wolf hunters in 1156, and two years later he paid out 29 shillings for the services of wolf hunters in Buckinghamshire and Bedfordshire, as well as 5 shillings and 6 pence to two wolf hunters who operated across Nottinghamshire and Derbyshire.[28] However, this does not necessarily mean that wolves were present in such areas – in some cases, wolf hunters may have been allocated to prevent wolves from re-entering. Salaried wolf hunters were stationed in Somerset, for example, despite there being no contemporary records of the animals in this county.[29] Nonetheless, the general distribution of hunting licences suggests that wolves were abundant across the north and west but scarce in the south,[30] reflecting the place-name evidence which indicates that wolves were much rarer in the south by the time of the Norman Conquest.

One of the most famous wolf hunters employed by the Crown was a baron called Sir Peter Corbet. Towards the end of the thirteenth century, Corbet was commissioned by Edward I to 'take and destroy all the wolves he [could] find in the forests, parks, and other places' of Worcestershire, Herefordshire, Gloucestershire, Shropshire, and Staffordshire.[31] Corbet was expressly given free rein to do whatever he felt necessary to kill all of the wolves in these counties – a letter written by Edward I to 'all Bailiffs' in May 1281 states that Corbet 'may take wolves with his men, dogs, and engines [traps], and may destroy them by all methods that may seem to him expedient', and commands the bailiffs to provide any assistance necessary to help Corbet in his task.[32] Special dispensation to hunt in the king's Forests was afforded to wolf hunters: Edward I allowed a man named John Giffard to hunt wolves in the king's Forests using both nets and dogs. Giffard was even excused if his wolf-hunting hounds accidentally broke free and attacked the king's deer.[33] Edward was evidently so concerned with eradicating wolves that he would happily sacrifice a deer in pursuit of his goal.

Kings also cleverly framed the right to hunt wolves within royal Forests – a useful service which rid them of their lupine competitors for the deer – as a reward for loyalty or service (or as bribery to ensure it – in the thirteenth century, hunting licences were given out to appease the barons who revolted against the Crown during the First Barons' War of 1215–17).[34] In the early twelfth century, Henry I granted a nobleman named Walter de Beauchamp the right to hunt wolves in all royal Forests in Worcestershire and 'to make traps round about his park for capturing wolves'.[35] Later that century, Henry II gave custodianship of Edinburgh Castle to a man named Roger de Stuteville,[36] along with licence to hunt wolves with dogs in Yorkshire and Northumberland.[37] King John subsequently allowed a man called William Brewer to hunt hare, fox, wildcat, and wolf 'throughout the king's lands',[38] and at the end of the thirteenth century Edward I granted one Richard Talbot 'licence, during pleasure' to 'hunt and take with his own hounds the fox, the cat, the wolf and the hare throughout the king's forest of Dene [Dean], and to take by nets or in any easier way the wolf there'.[39] Edward I also permitted a man named Roger de Mortuo Mari to hunt foxes, hares, wolves, badgers, and wildcats with dogs in the nearby Forests of Shropshire and Staffordshire,[40] while Edward III (r.1327–77) rewarded his surgeon, William de Holm, for his service with permission 'to hunt stags, hinds, bucks, does, goats, hares and wolves in all the king's forests, chaces, parks and warrens, and to carry away all that he shall take therein'.[41]

These records suggest that wolf hunting was sometimes enjoyed as a thrilling pastime alongside the hunting of a variety of other animals. Paradoxically, despite being reviled as a pest, it is possible that the wolf was simultaneously admired for its 'strength, speed, strong scent and self-confidence', with the danger, courage, complexity and skill involved in hunting the animal rendering it 'an interesting and challenging quarry'.[42] To successfully hunt a wolf required 'advance preparation and an economic insouciance beyond the possibilities of the average yeoman',[43] signifying status not because of the nobility of the animal, as was the case with deer, but because it required money and time which few had available. On the other hand, it could simply be that being permitted to hunt within the king's Forests was a privilege in itself, whatever the quarry.[44]

In a text claiming to record the Forest Laws in place during the reign of King Cnut of England (r.1016–35) – but which was in fact a forgery composed in the mid to late twelfth century during the reign of Henry II,[45] and likely reflects the Forest Laws in place when it was composed[46] – wolves were not considered 'beasts of venery' (game animals). Hence, killing them was not subject to punishment unless it occurred within Forest limits, in which case it was penalized 'gently'.[47] In comparison, the unlicenced killing and/or chasing of deer, hares, and rabbits within Forests was severely punished, as was cutting down trees.[48] The crime was not killing the wolf, to which little value was attached, but violating the king's exclusive hunting rights.[49] Similarly, a twelfth-century Welsh law proclaimed that 'no legal worth [was] fixed' on wolves, so that anyone was 'free […] to slay' them.[50]

The Norman kings also rewarded wolf killing through bounties, which increased in value as time went on and wolves presumably became rarer and more difficult to find and kill. During the reign of Henry II, 10 pence was paid out for the heads of three wolves in Hereford. But by the reign of King John, a dead wolf could be worth as much as 5 shillings – several hundred pounds in today's money.[51] A bounty system was also in place in Scotland from at least the early fifteenth century. In 1428, James I (r.1406–37) decreed that every baron in Scotland was to hunt and kill wolf pups in the spring, as well as on a minimum of four occasions during the rest of the year and whenever a wolf was seen within his baronry. To help with these tasks each baron was entitled to call upon his tenants, who could not refuse to join the hunt when called upon but who would be rewarded with 2 shillings for each head that they brought back. Lest anyone get any untoward ideas about the king's sovereign hunting rights, however, 'no man' could 'seek the wolves with shot' outside of the prescribed hunting seasons.[52] Thirty years later, James II (r.1437–60) ordained that local magistrates must gather the 'country folk' together three times a year between April and August, to kill wolf pups before they matured. The reward for any lucky wolf-slayer was 6 pence from the lord, bailie, or baron, as well as a penny from each household in the parish where the wolf was killed, presumably because it was considered a public service which benefited the entire community. Throughout the year every person was to be prepared to hunt as soon as a wolf was spotted, and there was no excuse for absence.[53]

Another method of controlling wolves involved the removal of trees among which they could seek shelter. An entry in the English administrative records (known as the 'patent rolls') of 1281 records licence being given to a Gloucester abbot to remove underwood in the Forest of Dean, 'because the said underwood is unfit to support the king's deer, and because wolves and malefactors of the forest frequently repair to it and stay there, by reason of its density'.[54] Wolves appear to have been a problem in this area: it was in the same year that Richard Talbot and Roger de Mortuo Mari were granted licence to hunt wolves in the Forest of Dean, Shropshire, and Staffordshire,[55] while in 1282 the Forest eyre (court) fined two men who scavenged from a deer carcass which had been brought down by wolves.[56] Yet the wolf depredations in this area apparently only got worse over the years, despite the efforts of these hunters. In nearby Mathon Park in Malvern, the deer stock was said to have been almost entirely destroyed by wolves by the year 1287, apparently not long after the park was created.[57]

The Norman kings also granted land to the nobility on the basis that the grateful new lord was to rid the area of its wolves. Kings William I, John, Henry III (r.1216–72), Edward I, Edward II, Edward III, Henry IV (r.1399–1413), Henry VI (r.1422–61 and 1470–1), and possibly others each parcelled out land all over England on the condition that it would soon be wolf-free, though in later years this may have become more ceremonial than practical as wolf numbers dwindled.[58] This system of granting land in exchange for the removal of wolves was also brought over to Ireland by the Norman invaders. In 1185 Henry II's son, John, who had been declared 'Lord of Ireland', granted his chamberlain land near Waterford, where he was permitted to hunt deer, boar, hare, rabbit, and wolf.[59]

The fate of the abundant wolf corpses that resulted from these efforts is unclear. Only a few fragments of wolf skeletons have been recovered from British and Irish archaeological sites dating to the Norman period. Foot and limb bones have been found in Wexford and Waterford, the former perhaps from a wolf killed by the garrison of a ringfort for its pelt, while the latter may have the product of trade.[60] In Rattray, Aberdeenshire, wolf bones have been found on the site of a former castle along with the remains of red, roe and fallow deer, wild boar, hare, and rabbit, which all may have been hunted in a reserve nearby.[61] A single wolf bone dating to the thirteenth century has been found in the Forest of Rockingham,[62] a large jawbone discovered in a medieval ditch in Northamptonshire has been 'tentatively' attributed to a wolf rather than a dog due to its large size,[63] and a wolf skeleton recovered from Helsfell Cave in Kendal has been radiocarbon dated to 1139–97.[64] Roe deer bones from the thirteenth century or earlier found in Rawthey Cave in Yorkshire appear to have been accumulated by wolves, suggesting that they were present in Cumbria at least up until the year 1300.[65] In 2024, canid bones found in the uncatalogued archive of English Heritage were the subject of excited speculation that they could have belonged to England's 'last wolf'. These remains were uncovered in Fountains Abbey in Yorkshire in the 1980s, along with other objects which may have been discarded when the abbey was disbanded during the reign of Henry VIII (r.1509–47).[66]

The dearth of surviving wolf remains in archaeological contexts suggests that the unfortunate victims of the extermination campaign were unceremoniously dumped in the wilderness, or perhaps burned. Though the odd foot or organ may have been salvaged first, for medicinal purposes. In his hunting treatise *The Master of Game*, Edward, Duke of York (c.1373–1415) notes that the wolf's front right foot could be used to cure 'the evil of the breast and for the botches (sores) which come to swine under the shoulder', and that 'the liver of the wolf dried is good for a man's liver'.[67] A fourteenth- or fifteenth-century text written in Middle English verse also gives a magico-amuletic usage of a wolf's tooth which, if accompanied by marigold, could not only protect against all conceivable dangers but could stop people from speaking anything but kind words to the amulet bearer, and even reveal the identity of a thief within one's dreams.[68] The wolf's tongue was far less useful, believed to be so venomous that it could kill any wolf foolish enough to lick its own wounds.[69]

In comparison to the squirrel and mustelid pelts that dominated the international fur trade, wolf fur was generally not very desirable.[70] It was considered coarse and unpleasant, disdainfully described by Edward, Duke of York as 'stinketh ever unless it be well tawed' (treated to turn the skin to leather). Edward does concede that the fur is 'warm to make cuffs or pilches (pelisses)',[71] however, and the skins could apparently fetch a small price. A thirteenth-century Welsh legal text values wolfskins at 8 pence (only a third of the value of a marten skin, despite the difference in size),[72] and records from 1430 indicate that 23½ pence was collected at the port in Aberdeen for wolf and rabbit skins.[73] The wolf foot bones found in Waterford and Wexford feature cut marks that indicate skinning which, along with the absence of any other bones, suggest that they may have been attached to pelts,[74] possibly indicating that wolfskins featured as a small part of the Irish fur trade.[75]

One of the last records of wolves in England describes monks in Whitby paying 10 shillings and 9 pence for 'tawing 14 wolfskins' in the last few years of the fourteenth century. These skins were perhaps the products of hunting or trapping locally,[76] and may suggest that these pelts had value to the monks or to someone to whom the monks could sell. Wolf fur may have been used as a cheap source of warm clothing or trim for the lower classes, who by law were only allowed to wear furs belonging to native species such as beaver, weasel, and otter, with the soft and exotic furs from sable, stoat, leopard, lynx, civet, and genet restricted to the upper classes.[77]

The Rise of the Sheep

It was not only competition for deer that led to such concerted persecution of wolves, but the ever-increasing importance of domestic animals upon which wolves might prey. Livestock were one of the pillars of the economy in Anglo-Norman Britain and Ireland, contributing to local trade as well as overseas commerce. But wolves did not know the value of the livestock they killed.

Since wolves tend to target wild species unless prey populations are depleted, wolf depredation on domestic animals was probably mitigated somewhat by the large numbers of deer that stocked the Forests and parks which dotted the British and Irish landscapes.[78] Attacks were also actively prevented by bringing livestock indoors at night and protecting them with shepherds and dogs during the day.[79] Of much bigger concern were disease and attacks by hostile people, which accounted for far more losses.[80]

Nonetheless, wolves did target livestock from time to time. A young horse and several sheep were killed by wolves in the Peak District in the mid-thirteenth century, and two horses belonging to the Bishop of Winchester were killed by wolves in 1209. In Lancashire, seven calves were attacked by wolves between 1295 and 1296, and eight cattle were killed in the same area a decade later.[81] In his *Master of Game*, Edward, Duke of York describes how 'if a wolf come to a fold of sheep [...] he will slay them all before he begins to eat any of them',[82] perhaps suggesting that surplus killing (whereby wolves kill far more animals than they can eat in a single sitting) was also observed in England.

No loss was felt as keenly as the loss of a sheep. The wool trade grew exponentially in the High Middle Ages, such that from 1250 to 1350 it was '*the* backbone and driving force in the English medieval economy'.[83] Between 40,000 and 45,000 sacks of wool were exported from England annually during the thirteenth and fourteenth centuries,[84] and at least 12 million sheep were kept across the country by the early 1300s.[85] By the end of the 1500s, around one third of England's land area was dedicated to sheep farming.[86] Likewise, in thirteenth-century Ireland the numbers of sheep on Anglo-Norman manorial estates almost doubled, whose enterprising owners increased their flock sizes to take advantage of wool shortages caused by disease among English sheep.[87] An average of 4,000 to 4,500 sacks of wool may have been exported from Ireland during the peak of its wool trade in the late thirteenth century.[88]

Scotland's wool industry was also hugely productive. The early fourteenth century saw 5,700 sacks of wool exported each year, rising to a peak of 9,252 sacks – or 1.5 million kilograms of wool – by the 1370s, which was produced by more than 2 million sheep.[89] Sheep farming was also hugely important to the Welsh economy, and although the amount of wool exported is unknown,[90] the estates of one family alone produced more than 18,000 fleeces in just five years in the fourteenth century, many of which were exported to London.[91] Cloth became the big money spinner in Wales in the following century, which saw the value of woollen textiles exported abroad rising from £198 in the early 1430s (around £200,000 today), to £816 in 1495 (almost £862,000 today).[92]

It took enormous numbers of sheep to meet the demand for wool throughout Britain and Ireland, with England's wool being particularly highly prized. Landowners took advantage of the profitability of sheep and the low labour cost involved in keeping them, especially in the wake of drastic population decreases following the Black Death in the mid-1300s – the bishop of Winchester alone had as many as 35,000 sheep on his estate in 1369.[93] Single families could have hundreds or even several thousand sheep to their name.[94]

But it was the monks of Britain and Ireland's monasteries who dominated the wool trade. One of the biggest players was the Cistercians, a religious order that acquired huge swathes of land across Britain and Ireland and razed woodland to make way for their money-making animals. Their activities are best described as 'large-scale sheep-ranching',[95] with some of their herds being unfathomably vast. While the Scottish Cistercians maintained flocks numbering in the thousands, the Cistercians in Ireland owned more than 200,000 hectares of land at their peak and were major exporters of wool.[96] The Welsh Cistercians likewise had more than respectable flock sizes (the abbey at Margam on the south coast, for example, possessed more than 5,000 sheep in the late thirteenth century, and Neath Abbey only a few hundred less).[97] But it was the English Cistercians who took sheep-ranching to new heights. The fifteen Cistercian houses in Yorkshire alone owned a combined total of 90,000 sheep, with Fountains Abbey housing almost 20,000 on its million acres of land.[98] Fountains even purchased its own ship to transport the wool.[99]

It was a booming business for these Yorkshire monasteries, who raked in astronomical sums from their woolly land tenants. Kings grew rich on the taxes from the enormous amounts of wool being traded out of the country, which provided about a third of the Crown's income during the reign of Edward III and almost two thirds by the time of Henry V (r.1413–22).[100] This is not to mention their income from the growing English textile industry, which boomed between the thirteenth and fifteenth centuries and eventually replaced wool as the powerhouse of English trade.

It is no coincidence that sheep numbers grew while the wolf population fizzled out, although which was the catalyst is impossible to know. Perhaps it was the eradication of wolves already underway that facilitated England's transformation into 'the biggest sheep farm in the world' and 'the source of its finest wool'.[101] Or perhaps the eradication of wolves was so doggedly pursued to protect the livestock animals whose coats were so valuable[102] – while many factors affecting the success of sheep farming were beyond human control, wolf numbers were much more manageable than freak weather events or outbreaks of disease, even if far fewer sheep were lost to depredation. Whatever the catalyst, it is clear that large-scale sheep farming would have been far more difficult, expensive, and time consuming, not to mention less profitable, if significant numbers of wolves were roaming the country. Sheep would need to be penned and guarded at night, which would inflate labour costs and restrict time spent grazing, in turn hindering the production of meat, milk, and wool.[103]

In his *The Last Wolf: The Hidden Springs of Englishness*, Robert Winder describes wool as 'the foundation stone of all subsequent English history'[104] – a stone that could only be placed with the demise of the wolf. Not only did sheep farming irrevocably change the natural landscape, but wool 'fostered the growth of ports', 'introduced England to large-scale commerce', and 'financed the construction of fine churches, guild halls and granaries' all over the country, creating a wealth of riches so enormous that it ultimately provided England with 'the flood of capital that allowed it to dominate first the British Isles and then an immense overseas empire'.[105] As one fifteenth-century Newark wool merchant put it, 'I thank God, and ever shall, It is the sheepe that payed for all'.[106]

Wolf land became wool land, wolves erased from landscape and map alike. Countless place-names mutated from *wulf* to *wool*, such as Ulvedel (*ulv* from the Norse *ulf*, meaning 'wolf') in Yorkshire which became Wooldale,[107] and Woolpit in Suffolk which, contrary to its name, was not once a place for washing wool, but a 'Wolfpit' (a pit in which to trap wolves).[108]

Wolf haunts shrank, the once marginal areas to which they could retreat increasingly exploited by people from the twelfth century onwards. More and more woodland was felled to accommodate the massive flocks of sheep, to create arable land, and to provide the building material, charcoal, and fuel needed to meet the needs of a steadily increasing population. What little tree cover was left – and it is estimated that woodland was lost at an average of 17.5 acres a day between 1086 and 1350, leaving only 10% of England forested[109] – was often maintained for hunting. The animal populations within these areas were carefully managed and, as we have seen, wolves were driven out. Wales also lost a lot of its tree cover, which had dropped to less than 10% by 1600.[110] Legend has it that the Normans felled much Welsh woodland to facilitate their invasion, which was hindered by the natural defences that the trees provided. Edward I supposedly brought an army of almost 2,000 woodmen to remove the trees in his path to conquest,[111] though there is little evidence to support such stories.

The human headcount increased exponentially between the years 1000 and 1350, by which point the British population had tripled to reach more than 3.1 million.[112] But the good times did not last. Already weakened by the Great Famine of 1315–17, the population was then decimated by the arrival of the Black Death in the 1340s. Settlements and farmlands were emptied of people, which may have allowed woodland to partially regenerate.[113] But it is unlikely that this was enough to allow wolves to recover.[114] Instead, the small numbers that survived were probably drawn to remote and sparsely settled upland areas where they could avoid people as much as possible, particularly in central Wales, northwest Scotland, and the Peak District.[115]

Yet the isolation of these refuges was a double-edged sword. The wolves ensconced in such places were cut off from other pockets of the increasingly small and fragmented total population. They ended up in genetic bottlenecks,[116] the final nail in the coffin for the species in England and Wales. Although wolves might have been spotted in northern England from time to time, making occasional southerly forays from Scotland in the fifteenth century, they were no longer full-time residents.

Conditions in Scotland and Ireland allowed wolves to survive for a little longer. Scotland's human population was relatively small, at just one million people at most, although the large majority lived rurally.[117] The Irish population was not large either, numbering just under one million even after massive population growth in the twelfth to thirteenth centuries.[118]

Much Irish woodland remained intact throughout and beyond the medieval period despite increased demand for wood after the English invasion, and it was only during the extensive woodland clearances of the seventeenth century that tree cover declined significantly.[119] This woodland will have provided a refuge for wolves in a landscape that was otherwise increasingly being developed and agriculturalised. Coupled with the fact that systematic persecution of wolves did not take place in Ireland until after the medieval period had come to a close, the species was able to survive longer here than in England and Wales.[120] On top of all this, the weakening of Norman influence in the fourteenth century also appears to have led the Anglo-Norman population to become 'largely absorbed into the customs and culture' of the native Gaels, whose attitude towards wolves was less hostile than that of the Normans.[121]

A lot of forest in Scotland had already been cleared prior to the eleventh century, and lowland tree coverage continued to shrink as the medieval period went on.[122] Nonetheless, there were still significant expanses of relatively undisturbed wooded uplands in the northwest which wolves could retreat to and, much of which was inaccessible to people.[123] Such places certainly appear to have had a reputation for containing numerous wolves, though whether these accounts are accurate is another matter. A map of Britain supposedly made by Edward II during the early fourteenth century depicts a wolf atop a Sutherland mountain, above which there is written in Latin 'here there are abundant wolves'.[124]

The Symbolic Wolf

Wolves and sheep were considered natural enemies, a relationship based partly in real-world behaviour but which was heavily reinforced by religious metaphorising. Even a harp string made of wolf's gut placed in amongst those made from sheep guts was said to 'corrupt' and 'destroy' the rest of the strings, according to thirteenth-century scholar Bartholomaeus Anglicus.[125]

The bestiaries that circulated in late medieval Britain and Ireland affirmed the inimical relationship between sheep and wolves in both literal and theological terms. A highly popular genre widely enjoyed by monastic and secular audiences alike, bestiaries were works of natural history which described the characteristics of various animals (more or less dubiously by the standards of modern science) and interpreted their behaviour allegorically to reveal religious truths. The wolf was a firm fixture of these texts from the twelfth century onwards. Described as a ferocious predator with an insatiable appetite, this cunning animal was said to have 'evil breath' which, if it was to successfully steal a sheep from a fold, forced it to stay upwind to avoid detection by guard dogs. The wolf also knew that it must be stealthy to sneak in and out of the sheep pen unnoticed, and was said to bite its own foot in punishment if it stepped on a twig and made a noise. It had eyes that shone like lamps in the dark, and the hair at the tip of its tail was a key ingredient in love potions. But it would not give up this prize easily: the wolf who feared that it was about to be captured would swiftly bite off the desired tuft which, being potent only if plucked from a live animal, was a decided middle finger to their pursuer.

Wolves also had the power to steal a person's speech, according to the bestiaries. Whoever this fate befell would be forced to take off all their clothes and bang two stones together to regain their voice and drive the wolf away. These stones represented the apostles or saints, to whom the person appealed to be forgiven for their sins, while the removal of the clothing signified baptism and the person's rebirth as a child of God, saved from the wolfish devil by their faith. Just as the wolf hunted and devoured sheep, this lupine demon was said to treat humankind as prey, stalking faithful Christian sheep in the 'fold' of the Church, looking for souls to devour. The wolf's shining eyes represented the works of the devil which, to the foolish, appeared beautiful

Figure 4.2. Illumination from a thirteenth-century manuscript produced in York (St John's College MS 61), showing a wolf stalking towards a flock of sheep. By permission of the President and Fellows of St John's College, Oxford.

and enticing, and it was unable to turn its head to look behind it, just as the devil is unable to 'turn back' and repent for his sins.[126]

Gloriously colourful images accompanying the bestiary texts were used to emphasise the wolf's predatory nature and its adversarial relationship with sheep, further reinforcing the allegory of the lupine devil stalking sheep in Christ's flock. Often the wolf in these illustrations is depicted creeping towards its ovine victims, but it is sometimes seen post-attack, running off with its prey as dismayed shepherds and guard dogs look on. In some cases, the wolves look more-or-less true to life. Although the individual in one thirteenth-century manuscript produced in York is decidedly cartoon-like (Figure 4.2), it stalks towards a flock of sheep with its head down and tail tucked in a recognisably lupine manner,[127] perhaps suggesting knowledge of the animal on the part of the artist. Similarly, while the wolf in one early fourteenth-century bestiary manuscript is exaggeratedly large, it is shaded in grey with pointed ears, a fuzzy tail, and highly detailed legs with a distinctive canine shape.[128]

Some, on the other hand, can be amusingly unrealistic. An early thirteenth-century manuscript possibly compiled in Durham shows what can only be described as a wolf on steroids, with a very chunky torso (perhaps in an attempt to illustrate the text's claim that the wolf has a strong chest but weak hind legs) and a grimacing face, which contrasts jarringly with a generally more accurately portrayed rear end (Figure 4.3). A fifteenth-century manuscript known as the 'Bestiary of Ann Walsh' depicts a very odd-looking blue wolf with short front

Figure 4.3. Illumination from a thirteenth-century manuscript produced in Durham, showing a large wolf at the entrance of a sheepfold. From the British Library Collection, Royal MS 12 C. xix.

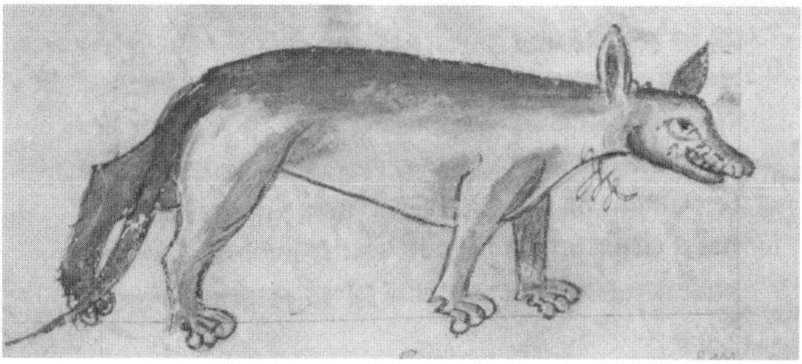

Figure 4.4. Illustration from the fifteenth-century 'Bestiary of Ann Walsh', depicting a somewhat misshapen wolf. Royal Danish Library, GKS 1633 4º: Bestiarius, f. 16v.

legs (Figure 4.4), a misplaced ear, a pointed tail which is so long that it drags along the ground, and an odd grin which makes it look more vulpine than lupine, although its back legs are far more realistic, if a little chunky. Either this artist was not very good at drawing, or they were somewhat confused about the appearance of wolves – perhaps because they had never seen one of these increasingly rare animals in person.

The wolf fared little better in the animal fables that circulated widely during this period. These classic tales attributed to Aesop were very popular, and were retold by a variety of figures including twelfth-century Anglo-Norman poet Marie de France, fourteenth- to fifteenth-century English monk John Lydgate, and fifteenth-century Scottish poet Robert

Henryson.[129] Wolves are invariably greedy, evil, and wily in these stories. They find any excuse to satisfy their voracious appetites and are decidedly unable to change their ways. One particularly popular tale was the fable of 'The Wolf and the Lamb', in which a wolf accuses a lamb of various crimes to justify its murder and consumption. In 'The Priest and the Wolf', meanwhile, a priest attempts to teach the wolf the alphabet, but by the letter 'C' the wolf's mind has strayed to its favourite sheepy snack.

Greedy wolves also feature in the beast epics (poems featuring animal characters which satirised human society) that circulated in medieval Europe, the most popular being the twelfth-century stories of a cunning fox named Reynard, who is often at odds with a rather less intelligent wolf named Ysengrimus. Entertaining tales of this fierce, cartoonish rivalry spread across medieval Europe, reaching England in the form of the late thirteenth-century poem *Vox and Wolf*, as well as oral stories which were likely imported from the Continent in earlier years.[130] In this poem the fox tricks the wolf into descending into a well, on the promise of finding a paradisical land with plentiful food of the ovine variety at the bottom. The chagrined wolf manages to escape when a friar pulls up the bucket, who declares that 'the devil is in the pit!' and promptly attacks it.[131]

Another story depicts sinful people being punished by God through transformation into wolves. This tale, related by Gerald of Wales (c.1146–1223) in his *Topographia Hibernica* (*Topography of Ireland*), is a miraculous account of two werewolves who encounter a travelling priest on the outskirts of Meath. While the priest is resting by the fireside, a wolf approaches and explains that he and his people were cursed by a saint named Natalis, so that every seven years one man and one woman had to take the form of a wolf and become exiles from humankind.[132] This story was perhaps written to emphasise the uncivilised nature of the Irish, their depraved and irreligious 'wolfishness', to justify Ireland's invasion by the Normans who would 'reform' this 'barbaric' country.[133] It is unclear whether literary portrayals such as these negatively affected the treatment of wolves in the real world, though they certainly will not have helped people to accept this difficult-to-live-with predator. The bestiaries in particular reaffirmed the wolf's status as an evil enemy of humankind, both literally and spiritually. Just as in previous centuries, the identification of Christians with sheep and of the devil with the wolf may have led to notions that the animal was dangerous to humans.

Wolves retained a reputation for eating human flesh (a reputation possibly based in reality, particularly when the Black Death created an abundance of corpses),[134] and were said to enjoy killing people. Bartholomaeus Anglicus described the wolf as an 'evil beast' which 'loveth well to play with a child' before killing and eating it, like a cat that plays with a mouse before ending its life. Not only this, but wolves could bear a grudge: if a person threw stones at a wolf, so Bartholo maeus says, the victim would seek retribution, killing the perpetrator if the stone hurt it or injuring them if it did not.[135]

Harking back to the Beast of Battle topos of Old English literature, Edward, Duke of York suggests that wolves gained a taste for human flesh after consuming corpses left in the wake of battles, also noting that old or toothless wolves would attack children because they were unable to bring down any other prey. Such beasts were highly dangerous because, so Edward writes, 'man's flesh is so savoury and so pleasant that when they have taken to man's flesh they will never eat the flesh of other beasts, though they should die of hunger', with reports of 'many men see[ing] them leave the sheep they have taken and eat the shepherd'. They are cunning animals who can sneak up on a person and, so Edward says, 'have a holding upon him before the man can see them'. Even a person who sees a wolf coming will have 'great difficulty' in escaping 'being taken and slain'. And even if a would-be victim escaped from the wolf's maw, if bitten 'he will scarcely get well, for their biting is wonderfully venomous on account of the toads they have eaten [...] and also on account of their madness'.[136] This latter description sounds suspiciously like a wolf infected with rabies. This disease is the culprit behind the vast majority of wolf attacks on people today and, most likely, throughout history,[137] including a wolf that reportedly attacked and bit more than twenty people in Carmarthen in 1166.[138]

Rabies may also lie behind the rising belief in werewolves at this time – one twelfth- to thirteenth-century English lawyer named Gervase of Tilbury (c.1150–1220) described lycanthropy as a 'daily occurrence among the people of our country: the course of human destiny is such that certain men change into wolves according to the cycles of the moon'.[139] Wolves with rabies act noticeably differently than uninfected wolves, being violent and far more dangerous, their mouths emitting foam and strange sounds. A person bitten by such a creature would also start to 'transform', becoming aggressive and making animalistic noises, perhaps even emitting lupine howls.[140]

Any extraordinary wolf – whether it killed an exceptional number of livestock, attacked people, or just behaved abnormally – could be explained away with werewolfery.[141]

As today, environmental factors may also have led to wolf attacks. It has been suggested that a record of wolves having 'killed many people' in 1420 in an Irish annal of Connacht may be attributable to desperation during a particularly cold winter, if not rabies.[142] An Irish annal of Clonmacnoise, meanwhile, records that a blind person was 'killed by wolves' in 1137,[143] perhaps an opportunistic attack on a vulnerable man which was noted in the annals because of its rarity.

Many stories of wolf attacks in medieval Britain are highly suspect. One such tale is told by English writer Walter Map (1130–c.1210) in his collection of anecdotes of Britain, *De nugis curialium* (*Of the Trifles of Courtiers*), a cautionary tale in which a man is attacked by two wolves in Wales after the wife of his host, resentful at their guest's intrusion, insults him and prompts him to leave. The unfortunate man manages to kill one of the wolves, but the woman's husband is too late to save him from a second.[144] An undatable story related by eighteenth- to nineteenth-century topographer William Hutchinson (1732–1814), which plays upon ideas of the medieval period as a wolf-infested wasteland,[145] describes a 'traditionary story of great antiquity' which relates how a place called Wotobank in Cumbria was so named after a woman was torn apart by wolves there while out hunting. Her husband, so the tale goes, cried 'Wo to this bank!', and the place has borne the name Wotobank ever since.[146] The name is in fact Wodow Bank, and is derived from the Old English *weald* ('forest') and *hoh* ('mound'), to which 'bank' was added some time later.[147]

Given the wolf's reputation for rapine, whether deserved or not, it is no wonder that this 'wonderfully wily' and cunning creature, deemed 'more false than any other beast', was described by Edward, Duke of York as hated universally and unwelcome everywhere, so that 'every man that seeth him chaseth him away and crieth after him'.[148] Nowhere is this hatred more evident than in a passionately anti-wolf Gaelic poem written in the fifteenth century by a man named Giolla Críost Táilliúr. The poet describes how the wolves which 'abound in each leafy glen' throughout Scotland 'devour horses and cattle and sheep' and 'wreak havoc on the people of the king', 'since they are incapable of being peaceful' and are allied with Satan. Táilliúr wishes nothing more than to see that 'their pernicious heads are slaughtered', so he calls upon God and Christ to 'send destruction', to 'burn' and 'curse' them, 'cleft their heads' and put them on spikes, 'lop away that misshapen brood', and shower 'murrain, rabies, cancer, [and] poison' upon the 'snarling ugly grey' beasts. He longs to hear 'deerhounds tearing the brutes asunder', praises a man who has already put 'many a cursed head on a stake', and enjoins him to continue with this campaign of destruction.[149]

Despite being an emblem of sin, violence, and the devil, in certain contexts wolves and wolfish violence appear to have been considered admirable or inspiring, as was the case in the early medieval period.[150] Though not appearing nearly as frequently as animals considered especially noble, like lions and eagles,[151] wolves were sometimes featured in the heraldry and emblems of aristocratic families in Britain.[152] Heraldry was an expression of identity, and it could have served as an indication of the bearer's power, courage, wildness, and predatory nature, perhaps in relation to the importance of hunting and hunting prowess to the nobility.[153]

The wolf often appears on the arms of families with a wolfish name. Two white wolves stride across a red background on the coat of a thirteenth-century Anglo-Norman nobleman called Nicholas le Lou (Nicholas the Wolf),[154] while the wolf also featured on the heraldry of families from places such as Wolverton and Wolverstone. Thirteenth-century seals of the Welsh Luvel family likewise depict a wolf, in recognition of their lupine name.

Others may have had wolves on their arms as a reflection of their personality. One of the best known is Hugh d'Avranches (c.1047–1101), a Norman-born Earl of Chester who was nicknamed Hugh Lupus (Hugh the Wolf), perhaps because of his prowess in battle. Hugh's heraldic device and that of his son featured wolves, as did the arms of a thirteenth-century earl of Chester named Ranulf (the animal has since appeared on some coats of arms for the city of Chester).[155] Three wolf heads also appear on the crest of the Louvell family, originally the lords of Cary Castle in Somerset but who settled in Scotland in the thirteenth century. The Louvell family name means 'little wolf' which, so various legends go, derives from one early ancestor or another from the eleventh or twelfth centuries, who was nicknamed either *Lupellus* ('Young Wolf') or simply *Lupus* ('Wolf') for their fierce temper and/or inhumane actions. Though this adopted family name softened to 'Louvell', the rest of the family were said to share their ancestor's terrible temper.[156]

Some heraldic wolves are said to be attributable to family myths about a wolf-slaying ancestor. Three wolf heads are depicted on the arms of Clan Robertson, whose lupine heraldry (and land) is said to be related to their founding member's purported proclivity for wolf hunting, such that he supposedly cleared the entire Atholl area of the Scottish Highlands of wolves.[157] The Anglo-Norman Engaine family, who were granted land by various English kings in exchange for hunting wolves, apparently had such a reputation for slaying wolves that the family seal depicted a running wolf surrounded by bits of broken spears and an axe head.[158]

The Last Wolves in England and Wales

The efforts of such wolf hunters paid off.

By the 1300s wolves were dwindling in number across England and Wales. Habitat loss, depletion of (and increased human control over) their prey,[159] and extreme persecution restricted them to increasingly remote areas.[160] In the first half of the fourteenth century, chronicler Ranulf Higden (c.1280–1363/4) wrote that England housed sheep that produced 'good wool' and 'many harts and wild beasts', but 'few wolves'.[161] In the year 1300 the rector of a church in London was brought before the mayor's court for the crime of receiving 'four putrid wolves' sent in a barrel from overseas, presumably because he could not find any closer to home.[162] A century later, when Edward, Duke of York wrote his *Master of Game*, wolf hunting was apparently no longer undertaken in England. Edward's text was a translation and adaptation of a highly influential hunting manual called the *Livre de chasse*, written by renowned hunter and French nobleman Gaston Phébus. But Edward tellingly neglected to translate Phébus's instructions on wolf hunting for his English audience,[163] evidently considering such information irrelevant for his readers, presumably because wolves were absent throughout much – if not all – of England. In fact, while Edward describes the wolf as 'a common beast', he states that there are 'few men *beyond the sea*', not in England, 'that have not seen some of them'.[164]

Wolves had probably disappeared from Wales and England by the fifteenth century.[165] The last clingers-on likely met their ends frightened, alone, and far away from people, but that did not stop local legends springing up about the 'last wolves' killed in England, Wales, or specific regions of the two. The New Forest, Whitby (Yorkshire),

Bolton Priority (Yorkshire), Wormhill (Derbyshire), and even the very unlikely locale of Cornwall are just some of the places that claim the dubious honour of being where England's 'last wolves' were slaughtered.

One of the best-known and most fanciful legends relates the demise of the 'last wolf in England', who met his end at Humphrey Head on the southern coast of the Lake District. First penned by a man named John Briggs in a letter written in the 1820s, who supposedly heard the tale – which apparently occurred during 'a remote period' of English history – from his uncle, the story was later put into verse by an anonymous poet in the 1850s.[166] According to this tale, a local landowner named Sir Edgar Harrington had sworn to kill this 'last wolf', which had been making a meal of the local sheep on forays from its coastal haunt. So determined was Sir Edgar to get rid of this troublesome wolf that he promised not only the hand of his niece, Adela, but half of his land to whomever accomplished the task. Two knights competed for the honour, one of whom happened to be Sir Edgar's long-lost son, John, in disguise, with whom Adela was in love. After an epic chase across mountains, forests, glens, hills, and crags, the wolf happens upon Adela. But just as the beast advances with 'glistening teeth', John appears and rescues the damsel in distress. Sir Edgar welcomes his long-lost son with open arms, and John and Adela marry in a nearby cave.[167]

John Harrington was a real person who lived between 1281 and 1347, though his father's name was not Edgar. The supposed wolf that lies at the feet of John's effigy on his tomb in Cartmel Priory is little proof of the legend's truth (the stone animal does not look especially lupine, and even Briggs' uncle acknowledges that, 'without a word of inscription, [the] monument remains to puzzle the fertile brains of modern antiquarians'),[168] but perhaps may have been the inspiration for the story.

Legends of the 'last wolf' killed in Wales betray their fancifulness by placing the event far later than the animal likely disappeared. In his *Lost Beasts of Britain*, Anthony Dent describes six places 'reputed by oral tradition to have harboured the last wolf in Wales'. Two of these places have wolfish names (Welsh *blaidd/bleidd*) which, though likely due to a former abundance of wolves in these areas, may have inspired the telling of such tales – at a bog in Y Bannau Brycheiniog (the Brecon Beacons) called Gwernblaedda, a 'last wolf' was supposedly slain at an unknown date, while another 'last wolf' was killed during the reign of either Elizabeth I or James I on a farm known as

Ffos-y-bleiddiad in Ceredigion. At Usk in Monmouthshire, meanwhile, the paw of the last wolf in the nearby Forest of Goytre, which was apparently killed in 'Tudor times "or later"', could supposedly still be found nailed to an oak tree as late as 1928. Parish records purportedly describe the slaying of a 'last wolf' on a hill in Powys, also in the late Tudor period, though other sources claim that the last wolf in Powys was killed in 1750. A Carmarthenshire valley is also said to be the site at which the last wolf in Wales was killed as recently as the late 1700s.[169] Yet more stories are relayed by Cledwyn Fychan in his history of wolves in Wales, *Galwad y Blaidd*, and by John Pollard in his book *Wolves and Werewolves*.[170]

The men who slaughtered these 'last wolves' likewise became legendary. Or, perhaps more accurately, men made themselves legendary by claiming to have killed a last wolf – as folklorists Jennifer Westwood and Jacqueline Simpson put it, 'heroes need slayworthy opponents'.[171] A Northamptonshire folk story tells of the brave exploits of a man named Jack of Batsaddle, whose life was tragically cut short after he killed the last wolf (and/or boar) in England. Ironically, Jack survived this dangerous encounter only to kick the bucket when he drank some water from a nearby stream, which was so cold that it killed him. This tale is attached to a supposed effigy of the wolf-slayer in a church in Orlingbury, which in fact probably belongs to a knight named John de Withmayle, who was memorialised not for any such feats of daring but for donating land to the church.[172] John of Gaunt, son of Edward III and originator of the House of Lancaster, is also credited with killing the last wolf in England, or at least Yorkshire, in the late fourteenth century.[173] Apparently he did not do a very thorough job – in the late fifteenth century, a man supposedly killed such an extraordinary number of wolves in Teesdale Forest that he was said to eat them, earning him the nickname 'Roast Wolf Ambrose'.[174]

Wherever and whenever the last wolves on English and Welsh soil took their final breaths, their extinction marked a turning point in the history of the landscape. The natural world had finally been brought to submission, unspoiled by these irritating, foul, evil, and wild beasts, whose pawprints had long blotted the landscape. Humans could now call themselves the only apex predator left in England and Wales, free to farm their sheep and hunt their deer unhindered.

William the Conqueror's vision had been realised.

CHAPTER 5

'Enlightenment': The Final Purge

You walk among the collection of exotic animals confined in the Tower. Your footsteps ring as you tread across the stone floor, contemplating another miserable elephant whose trunk stretches towards the light beyond its bars, desperate for freedom, desperate to eat something other than the wine and meat he is forced to consume.

This place feels distinctly more like a prison than a zoo. It smells like shit and fear.

Further down the row, a troop of monkeys clamour at their cage, the whites of their eyes gleaming as they screech.

You wonder who could possibly enjoy seeing these animals. The excitement of witnessing a lion in the flesh for the first time is surely not satiated by one who paces the length of his too-small enclosure, mangy and broken from boredom and stress.

The wall of cages stretches on and you quicken your pace, weary from gazing into the same sad eyes and broken souls behind every set of bars.

Then, at the end of the row, you finally find it. The only wolf in England.

He is unnaturally alone, skinny and lethargic, squashed against the back wall of his cramped cage, cowering in the dark. He has been baited and battered for a reaction enough times that he snarls the moment you appear, though he does not raise his tired head. Life is leaving him.

Far away, across the northern border and over the western sea, his wild cousins are equally frightened and desperate. Their freedom is little consolation – they are treated no better than their cousins in captivity, hunted with an almost deranged fervour. They are poisoned, trapped, tormented, and tailed at full pelt by men on horseback brandishing shining steel and blazing gunpowder. They run and run until their energy is spent. Exhaustion, dogs, and death snap at their heels. Families are separated, confused, terrified. But the forests are gone, and people are everywhere. The wolves have nowhere to go. They will be the last of their kind.

The Wolf in the North

Wolves were still clinging on in Scotland by the sixteenth century, though their numbers were almost certainly far less impressive than some sensationalist writers from the time would suggest.

Many authors in the sixteenth century had a penchant for depicting Scotland, especially the Highlands, as an 'untamed', wolf-infested wilderness at the far reaches of civilisation. In his 1527 *History of the Scottish People*, philosopher Hector Boece (1465–1536) described a 'great supply' of wolves in Scotland, who were 'most cruel towards our herds, and very hostile to our cattle', and were found throughout 'the entire realm' except, tellingly, 'where human habitation prevents it'.[1] Fifty years later, Scottish bishop John Lesley wrote that there were still 'very many' 'most cruel' wolves in Scotland, mostly in the north – a statement whose veracity is called into question by Lesley's claim that these animals attacked not only livestock but 'even men, especially women with children'.[2] William Harrison likewise depicted a glut of wolves in Scotland which, drawing on Boece when composing his contribution to *Holinshed's Chronicles of England, Scotland, and Ireland*, he described as 'most fierce and noisome to the herds and flocks in all parts of Scotland'. Strangely, Harrison goes on to note that in 'one parcel of Angus, called Glennors dale', wolves 'do no manner of hurt unto the domestic cattle, but prey only upon the wild'.[3] There is no explanation for these apparently harmless wolves.

In 1618, English poet John Taylor (1578–1653) depicted himself as an intrepid adventurer in an uninhabited wilderness as he travelled through Scotland on foot, describing how it 'was the space of twelve days [...] before I saw either house, corn field, or habitation for any creature, but deer, wild horses, wolves, and such like creatures, which made me doubt that I should never have seen a house again'.[4] Wild horses disappeared from Europe around the early Mesolithic, so any horses seen by Taylor were almost certainly owned by people,[5] suggesting that the landscape he traversed was not so sparsely populated as he would have his readers believe. It is also highly unlikely that he actually saw wolves in the wild. At most, he would have heard their howling.

Such assertions of a pestilential population of wolves throughout Scotland are belied by the complete absence of records of them in the southern lowlands at this time. Wolves appear to have been long since extinct in these areas – while ten wolves were killed in the Borders in the mid-1400s, after this point they disappear from the record and, presumably, from the landscape of southern Scotland. Large flocks of sheep moved into the newly wolf-free Borders, Holyrood Park, and Fife in the sixteenth and seventeenth centuries, after which there is no mention of lupine depredations in these areas.[6]

Even as early as 1570, wolves were apparently hard to find in Scotland. In a letter written to the Countess of Moray, one Alexander Clark expresses regret that he was unable to fulfil the countess's request for wolfskins, of which he 'could get na knowledge [...] at the present'.[7] The situation did not improve in the following years – in 1584, a Stirling merchant reneged on a contract to provide a French businessman with 600 wolfskins. Wolf pelts appear to have been held in greater esteem around this time, likely as a result of the animal's increasing rarity. A Perthshire laird possessed four wolfskins in his 'household garderobe' in 1604, while in 1661 the court of King Charles II (r.1649–51 as King of Scotland, 1660–85 as King of England) paid 2 ounces of silver for between 10 and 12 pelts.

Wolves had become 'exotic' enough that Scottish royalty also kept them in their menagerie collections. In 1598, a keeper of the animals at the Holyrood menagerie was paid 26 pounds, 13 shillings and 4 pence 'for the entertainment made by him upon two wolves' (the meaning of this is uncertain).[8] A century or so earlier, a man who brought a live wolf to King James IV (r.1488–1513) was given five shillings for his trouble, though it is unclear whether the animal was considered 'sufficiently uncommon to be preserved as [a] curiosit[y]' in the Royal Menagerie.[9]

Their scarcity may also lie behind the transformation of wolves from a beast unworthy of the hunt to a desirable object of the chase in the eyes of the Scottish elite. Although deer remained the most prized quarry, John Lesley described that 'the nobility [had] their most recreation in hunting with the sleuthhounds', with which they enjoyed pursuing 'the hare and the fox [...] or in the mountains the wolf, or the wildcat'.[10] Even kings and queens hunted wolves. Mary, Queen of Scots (r.1542–67) is famously said to have partaken in an extensive hunt in Atholl in the early 1560s. During the hunt, a staggering 2,000 Highlanders were apparently employed to drive a great herd of around the same number of red deer towards the hunting party. Several of the Highlanders were killed when the queen set one of her hounds on a wolf that was caught up in the chaos, causing the deer to stampede. In total 360 red deer and 5 wolves were killed, and although the latter were likely unintentional captures, having simply been driven to the slaughter-ground along with the deer, they were probably a 'welcome [...] addition' to the bag nonetheless.[11]

Hunting was not just the province of wealthy men and women, however, and continued to be enforced by landowners who wanted to keep wolf numbers down. In 1552, a lease for land in Glenisla stipulated that the tenants had to maintain a pack of dogs with which to hunt foxes and wolves whenever they were called upon to do so.[12] Likewise, the tenants of an estate in the Highlands were required to create weapons for wolf-hunting in 1621, under the threat of a fine if they did not comply.[13]

The notion that wolves were a widespread scourge in the north of Scotland goes hand in hand with the misguided idea that the Highlands were host to vast swathes of uninhabited, uncivilised wilderness. Victorian 'authorities', such as the notorious Sobieski-Stuart brothers,[14] liked to imagine the Scotland of days gone by as a pristine wilderness, with 'wild and desolate moor-land hills', 'great forests', and 'immense tract[s] of desert mountains utterly uninhabited, and unfrequented'. Such areas, so they said, 'served as preserves' where the 'savage race' of wolves 'increased at intervals to an alarming extent'. These forests were 'so dense, and infested by the rabid droves, that they were almost impassable', the wolves actively helping to maintain these frightening forests by serving as guardians of the wilderness frontier. The 'numbers and ravages' of these beasts 'were formidable',[15] and the only solution was to fell or burn the 'large tracts of forests' to 'expel the Wolves which there abounded'.[16] This sensationalised image of a wolf-ridden landscape was repeated by other Victorian scholars such as James Edmund Harting, whose chapter on the species in his

British Animals Extinct Within Historic Times is in turn still frequently treated as an authoritative account today.[17]

But the Sobieski-Stuarts were far from the first to mythologise the Highland wildernesses. Legends about the ancient Caledonian Forest had long since proliferated, having first been very vaguely recorded by Roman writers (who could not decide where, exactly, this supposedly vast forest had been), quite possibly to explain the Empire's fail ure to colonise northern Britain. Writers in the Early Modern period evidently enjoyed the notion of the great wilderness forest. Boece depicted an ancient woodland occupied by mythical white bulls with leonine manes, and English topographer William Camden (1551–1623) threw in 'witches and bears for good measure'.[18]

Such ideas about the wild, wolf-infested wastes of the north and their savage inhabitants remained popular, later becoming entangled with English perceptions of northern Scotland as a place that needed to be civilised and brought under control. A soldier who travelled to the Highlands in 1656 (as part of a military expedition to assert English control after Oliver Cromwell's (1599–1658) defeat of Scotland in the Anglo-Scottish war of 1650–2) wrote that the people of Strathnaver were 'almost as barbarous as cannibals',[19] while another described the most isolated and densely wooded parts of the Highlands as home to 'ravenous wolves with two legs'.[20] The wolf was symbol of a wild and untamed place and a people that both had to be subdued.

But in reality, by the early modern period much of Scotland's landscape had already been cleared of trees to make way for agriculture. It is estimated that a mere 4% of Scotland was wooded by the mid-1700s.[21] Scotland's southern uplands had already been taken over by sheep farming in the medieval period,[22] while the lowlands housed large flocks sometimes numbering in the thousands by the mid-1500s.[23] Large-scale introduction of sheep to the Highlands, meanwhile, had taken place by the end of the eighteenth century.[24]

Wood was thus a rare and precious commodity in Scotland, and much raw material had to be imported from abroad to meet demands in the absence of a local supply. Though small pockets of woodland that could sustain local people did survive, the remaining forests could not keep up with the levels of construction pursued during the reigns of James IV and James V (r.1513–42), and in 1503 the Scottish parliament declared 'the wood of Scotland' to be 'utterly destroyed'. Although they encouraged landowners to plant trees to remedy the situation,[25] little changed in the following 50 years, and in 1564, the Privy Council of Scotland considered it likely that woodland in the north would soon be entirely eradicated.[26]

The wolves that were clinging on in Scotland did not live in any great woodland wilderness, therefore, but in 'open or scrubby country', where they were forced to compete for space with livestock.[27] Unlike the elk, bear, beaver, and possibly lynx,[28] which had all disappeared by this time due to fragmentation of their woodland habitats and human disturbance of what little forest remained, wolves could survive without large extents of tree cover.[29] Forests were almost certainly not burnt to a cinder to expel wolves. Wood was too precious, and clearing the trees would scarcely help to get rid of wolves anyway.[30]

But though they could live without woodland, wolves could not so easily survive a lack of prey. Widespread hunting and increasing numbers of livestock to compete with saw red deer numbers plummet by the middle of the sixteenth century.[31] With few options left, wolves turned to preying on livestock – particularly the cattle that were moved to pasture on higher ground during the summer.[32] But livestock was fiercely protected, especially by subsistence farmers who would feel keenly the loss of a sheep or cow to a wolf, particularly in the harsh winter months.[33] The best that the dwindling wolf population could do was retreat to the Highlands (particularly the most northerly and remote county of Sutherland), following the remaining deer that sought sanctuary in these areas.[34]

Even here, they could not escape the persecution of humankind's pen. Sixteenth- to seventeenth-century cartographer Timothy Pont (c.1565–1615) put it relatively mildly when he demarcated part of Sutherland as an 'extreme wilderness' full of 'many wolves' on a hand-drawn map.[35] William Camden pulled fewer punches in his 1588 chronicle *Britannia*, where he proclaimed Sutherland to be 'haunted and annoyed by most cruel wolves', which 'in such violent rage not only set upon cattle to the exceeding great damage of the inhabitants, but also assail men with great danger, and not in this tract only, but in many other parts likewise of Scotland'.[36] Such sentiments were shared by another cartographer named Robert Gordon of Straloch (1580–1661), who depicted Sutherland as home to numerous 'violen[t]' and 'most rapacious wolves which here, prowling about wooded and pathless tracts, cause great loss of beasts and sometimes of men', having been 'driven from almost all the rest of the island'.[37]

There are also folkloric stories of wolves digging up graves, forcing communities to bury their dead on nearby islands (the best-known case being Handa, an island off the Sutherland coast, where the people of the mainland settlement of Eddrachillis supposedly buried their dead),[38] or within roundhouses (known as brochs). It was also allegedly customary in the Highland village of Atholl to use coffins made of multiple layers of stone to deter wolves from digging them up, whilc in the Assynt area of southwest Sutherland, the custom was apparently to build cairns for the same reason.[39]

Such stories are likely exaggerations based on perceptions of a remote 'wilderness' haunted by bloodthirsty, man-eating creatures, a product of the Victorian penchant for mythologising both the wolf and the past. While wolves are scavengers and food sources were scarcer during this time, as natural historian Jim Crumley notes, 'grave-robbing sounds too much like hard work to reach a source of carrion', and probably occurred only if 'the corpse was hastily buried in a shallow grave and without any kind of coffin'. Instead, he suggests that the practice of burying bodies on islands is 'more likely to have originated in communities where good grazing land was at a premium: no point in wasting it on a field of the dead'.[40]

The Last Wolf in Scotland

Although legends of Scotland's 'last wolves' imply that they survived in northern Britain up until the very late date of 1743, the non-mythological evidence indicates that wolves became extinct in Scotland somewhere around the first half of the 1600s. At least one was alive in 1621, when a substantial bounty of £6, 13 shillings, and 4 pence (around £1,575 today) was paid out for a wolf killed in Sutherland.[41] This large sum is likely indicative of the animal's rarity, as well as a fervent desire to pick off the last few survivors of its species. By 1684, Scottish physician Robert Sibbald (1641–1722) noted that Scotland 'lack[s] those wild and savage [animals] of other regions', including wolves, which had been 'extirpated from the island'.[42] In 1769, when Thomas Pennant 'travelled into almost every corner' of Scotland, he noted that he 'could not learn that there remained even the memory' of wolves even 'among the oldest people'.[43]

Legends of the 'last wolf' in Scotland (or one of its regions) are ubiquitous and sensationalist in equal measure.[44] They are indicative of little more than the seemingly inevitable mythologising that hounds these animals wherever they go, even pursuing their ghosts long after their corporeal disappearance.

According to the eighteenth- to nineteenth-century hunter William Scrope (1772–1852) – whose opinion of wolves and the unreliability of his testimony are both made clear by his descriptions of these animals as 'detested prowlers' and a 'rabid race' – the last wolf in Scotland was killed in the Sutherland valley of Glen Loth in the late seventeenth century. At this time, so Scrope says, it had been thought that 'the villainous race was extinct', but this 'last wolf' gave away its presence with its lonely howling and its attacks on sheep. After an unsuccessful search for the beast, an old wolf hunter named Polson took his son and a shepherd boy into 'the wild recesses' of Glen Loth to kill the creature. Polson soon tracks the wolf down to its rocky, cavernous den. The boys enter the den to find five or six wolf pups which they quickly begin to slaughter. But the pups' 'feeble howling' draws their mother, who 'rag[es] furiously at the cries of her young'. She dives into the hole before Polson can apprehend her, but as she desperately leaps in to save her young, the hunter fortuitously catches her by the tail and winds it around his arm. After a struggle, during which there is a now oft-quoted exchange – '"Father, what is keeping the light from us?" — "If the root of the tail breaks," replied he, "you will soon know that"' – Polson manages to stab the wolf with his hunting knife.[45] There is no mention of what happened to the father of the wolf pups. In 1924 a monument was erected by the Duke of Portland to commemorate this event; a small, unassuming stone which claims to 'mark the place near which [...] the last wolf in Sutherland was killed by the hunter, Polson, in or about the year 1700'.[46]

In a footnote to this tale, Scrope describes a similar story told by Scottish writer James Hogg (1770–1835) in his *Winter Evening Tales, Collected Among the Cottagers in the South of Scotland*. According to this tale, two men go out hunting in the hope of catching one of the 'few wild swine remaining' in the Highlands. They find 'a deep pit or cavern, which contained a large litter of fine half-grown pigs'. As in Scrope's tale, one of the men enters the den to dispatch the litter, but an adult boar soon arrives on the scene, furious that their young are being slain. So too does the boar attempt to dive into the hole, only for the second man to grab it by the tail, which he wraps around his hands. So too does the man in the cave wonder what is blocking the light, to which the other replies: 'Should te tail preak, you'll fin' tat.' So too does the man finally extinguish the beast with his dagger.[47]

Scrope asserts that Hogg had heard the story of Polson's wolf and altered it to 'make the tale his own'. In replacing the wolf for a boar, however, Scrope claims that Hogg 'has fallen into an error which lessens its probability', before going to great lengths to attempt to prove that his own story is true. Scrope's rebuttals include noting that the boar's tail 'is proverbially short, and of slender dimensions, and could hardly be grasped firmly by the hand'. He also suggests that 'a sow or boar also invariably roars out most lustily when seized or obstructed, and hence the person in Hogg's cavern must have known from such sounds the cause of obstruction of the light without further inquiry', whereas the wolf, he contends, 'defends himself in silence'.[48] It apparently never occurred to Scrope that the origin of both stories could be a folkloric tale (and in fact, an almost identical story is told of a she-wolf and her pups killed in Gwynedd (northwest Wales) in 1785, and there are several very similar stories from other locales in Scotland),[49] whose mythical nature is betrayed by its appearance as both a 'last wolf' and a 'last boar' story. The irony of his assertion that 'Polson's exploit [...] was a true one' appears to have been lost on Scrope.[50]

Many other 'last wolf' stories are likewise so similar that they appear to be the same legend simply adapted for each locale in which they are set.[51] Numerous 'last wolves' have, coincidentally, been dispatched by women with frying pans, including an old Highlander who 'brained' a wolf which was rude enough to interrupt her as she sat gossiping with a friend.[52]

Another woman was supposedly attacked by a wolf while walking through Argyllshire, and though she was eventually killed she did put up a worthy fight, using a knife to pierce the animal's heart and an apron to protect the arm that wielded it.[53] Elsewhere, a man supposedly killed the last wolf on the Isle of Skye by thrusting a deer bone down its throat after it attacked him, which is said to be memorialised on his clan coat of arms.[54]

Another story comes from a footnote in a late eighteenth-century travelogue of Scotland, which notes that a famous chieftain, Sir Ewen Cameron (1629–c.1719), 'is said to have killed the last wolf in Scotland, about the year 1680'.[55] The body of this wolf was supposedly on display in the London Museum before being sold in the early nineteenth century, described in the auction catalogue as 'Wolf – a noble animal in a large glass case. The last Wolf killed in Scotland by Sir E. Cameron'.[56] This, of all the stories of Scotland's 'last wolves', seems the most likely to have any truth to it (least of all for its lack of fantastical details) – if there is any truth to be found in such legends at all.[57]

The body, on the other hand, is yet another myth. When writer Adam Weymouth attempted to track down the history of this wolf, he discovered that although it is now lost, it was certainly not 'last'. The man who founded the London Museum in 1807, Edward Donovan (1768–1837), had purchased the wolf from one James Parkinson (1730–1813) in 1806, who in turn had bought it in 1786 from collector Ashton Lever (1729–1788). A zoologist named George Shaw (1751–1813) studied Lever's collection, publishing a catalogue and documentation of it in 1792.[58] In this book, Shaw described that the wolf was in fact not the last wild wolf in Scotland at all, but the stuffed remains of a tame wolf that had been Lever's pet. A few decades later, when Donovan was forced to sell his collection due to bankruptcy, he apparently 'turned his specimen into "the last Wolf" in the hope of a few extra quid'.[59]

The remnants of another 'last wolf', supposedly killed on the border between Caithness and Sutherland, were also the subject of a case of mistaken (or, rather, mis-sold) identity. A local family claimed to have killed this wolf in 1763, from whose corpse they had rather inexplicably retained only a jawbone. It is telling of the grand mythologising that surrounds 'last wolves' that this bone in fact belonged to a tiger.[60]

But it is the tallest tale that is most often repeated. According to this story, the last wolf that 'infested' the Moray coast at Findhorn had attacked a woman and her two children as they attempted to cross the mountains in the winter of 1743. Only the woman escaped with her life – the children were eaten by the 'black beast'. Having heard the woman's story, the chief of Clan MacIntosh ordered his people to assemble a hunting party to kill the animal. One member of this clan, a celebrated deerstalker called MacQueen, was 'looked for to take a lead in the enterprise', being a gigantic 'seven feet six inches in height' and 'possessed of gigantic strength and determined courage'. But on the morning of the hunt, MacQueen did not turn up. The chief waited impatiently until MacQueen finally appeared, nonchalantly feigning forgetfulness before dramatically producing the 'grim and bloody head of the monster' from the folds of his tartans. He regaled the company with a gripping tale of how he grappled with the beast and cut its throat, bringing the head back for good measure 'for fear he might come alive again; for they are very precarious creatures'.[61]

Despite its sensationalist nature, the year in which this story is set is frequently given as the date at which Scotland's wolf population became extinct, including by naturalists and other authorities. But it is almost certain that the story is a fabrication, not least because even some 82 or more years earlier, wolves were 'believed to be now all but extinct' in nearby Aberdeen and Banff, according to Scottish antiquarian Walter MacFarlane (d.1767).[62] Not to mention MacQueen's gigantism – his height conveniently matches the (mythologised) height of William Wallace[63] – as well as the wolf's apparent jet colour, which is extremely rare in Europe,[64] and the animal's highly irregular man-eating tendencies. It is also telling that a member of the MacQueen family is also said to have killed the last great auk in Britain.[65]

Although it is possible that there is a kernel of truth somewhere in this story, it has undoubtedly been buried under years and years of telling and retelling. Perhaps it was inspired by a coat of arms of the MacQueen family, which a book of heraldry published more than 20 years prior to MacQueen's supposed heroic feat describes as bearing three wolf heads.[66] It is no coincidence that stories about terrifying black man-eating 'last' wolves killed by heroic huntsmen are found not only throughout Britain but also across mainland Europe.[67] Tales of gigantic, child-eating black wolves say much less about wolves than they do about humans and our penchant for myth-making.[68]

The exact date of the wolf's extinction in Scotland is known to no one, but it is almost certainly later than the last recorded encounters with the species. The few survivors, with an ingrained fear of humans after years of persecution and perhaps first-hand experience of their ruthless campaign of lupicide, 'would have been almost impossible to find if they chose not to be found', as Jim Crumley puts it.[69] Dwindling numbers of wolves could have survived in remote regions of the country, undisturbed and unknown. It was the unviability of such fragmented populations, not the masterstroke of a gargantuan warrior in tartan, that ultimately led to their demise. As Crumley vociferously writes in his investigation of these legends, the true last wolf likely 'died old and alone in a cave, somewhere [...] far from the gaze of humankind'.[70]

No heroic hunters, no man-eating beasts.

Just a lost, lonely wolf, who howled across the glens but heard no reply.

The Last Wolves of the Emerald Isle

Wolves survived beyond the medieval period in Ireland in no small part because the Anglo-Norman colonisers of the fourteenth century were slowly but surely absorbed into the wolf-tolerant Gaelic culture.[71] But with the sixteenth century would come another English invasion and, along with it, a freshly imported hatred of wolves and a renewed campaign of lupicide.

Though the natives did hunt wolves, with a significant trade in wolfskins from Ireland evidencing a large population able to sustain losses of at least several hundred individuals each year during the 1500s,[72] the Irish were not at pains to eliminate the species. For them, wolves were simply part of the natural world – just one of many animals with which they shared the landscape.[73] As in previous centuries, they opted to protect their livestock within ringforts rather than killing wolves to prevent depredations.[74]

But the English took a rather different view of the 'great store' of wolves that inhabited the island.[75] In 1542, King Henry VIII (r.1509–47) declared himself King of Ireland to reassert English rule on the emerald isle, which had slowly been reclaimed by the Gaels since Henry II's conquest in the twelfth century. Land was confiscated from the natives and given to British settlers, Protestantism became the official religion, and English law was enforced, leading to many conflicts and rebellions over the following centuries which shaped the course of Irish history to this day.

Entwined with the politics and land-grabbing aspirations of the English was the belief that Ireland required 'civilising' and 'taming'. This aim was squarely at odds with the abundance of wolves that Ireland housed. To the English settlers, the wolves were a verminous threat which had to be eradicated if the anglicising and civilising of Ireland was to be successful.[76] It was also a justification for English intervention: the English had 'successfully "tamed" England's landscapes by eliminating wolves', and if the Irish couldn't do the same themselves, then the English would do it for them.[77] And they would claim the newly de-wolfed land as payment for their efforts.

The English colonial aspiration was also threatened by the presence of wolves in human form. Like the country itself, the English viewed the Irish – particularly those who resisted the invasion – as primitive, uncivilised, and wild, and they frequently likened the natives to the wolves that they saw as a scourge on the landscape. English poet Edmund Spenser (1552/3–1599) wrote in 1596 that the Irish 'make the wolf their gossip' (not suggesting that they talked to wolves, but that they were 'Godsiblings'[78]), an affinity so deep that the Irish could even take lupine form.[79] William Camden similarly noted that the Irish prayed for wolves and 'wish[ed] them well, and so they are not afraid to be hurt by them'.[80]

Rebels known as 'woodkernes', who launched counterattacks from their woodland haunts,[81] were perceived to be as much of a threat to the English colonial project as 'the devouring wolf', with one English politician recommending in 1610 that woodkerne and wolf alike should be hunted down.[82] Likewise, a 1598 tract described the native Irish as 'ravenous', 'savage', and 'unmerciful' wolves, who preyed upon the 'poor innocent' English 'as the wolves do upon sheep',[83] while an Irish garrison killed by the English in 1647 were said to have 'tails near a quarter of a yard long'.[84] In a 1642 political text titled 'Irelands Tragical Tyrannie [...] Wherein is Plainly and Truly Shown, what Cruelty Hath Possess the Irish Rebels Hearts', which details the 'savagery' of both the Irish rebels and wolves alike, the former are depicted attacking and butchering a woman, while the latter are seen killing and eating an innocent English family of fourteen.[85]

The wolfishness of Ireland and its uncivilised inhabitants, especially in comparison to the wolfless island to the east, was thus the perfect justification for the invasion and the slaughter of both people and animals. For the English, wolves were a symbol of a landscape that had not yet been brought under control so that human society could flourish. Ireland needed to be purged of wolves both human and animal. Woodkernes and wolves were hunted down and killed, bounties placed upon human and animal rebels alike.[86]

There are legends that the English army razed much of Ireland's remaining forest to smoke both types of 'wolves' out of their woodland 'lairs'. As in Scotland, however, there is little evidence for such assertions,[87] although woodland certainly offered habitat for wolves along with uncultivated mountainous regions, bogland, and areas of exposed limestone karst. Combined, these areas provided around 27,500km^2 of relative 'wilderness', which could have housed anywhere between 250 and 780 wolves.[88]

Tree cover did fall dramatically during the seventeenth century, but this was due to the expansion of agriculture, the granting of land to English settlers who cut down vast swathes of forests to make a profit, and felling being a condition of many land leases.[89] While it is estimated that woodland coverage in 1600 was around 12.5%, within a century this had been reduced to just 2%.[90] The presence of wildland which had not been modernised and optimised for human use was another sign of the country's supposedly uncivilised nature, and the English, 'as pioneers of the new scientific rationalism' of the early modern period, saw it as 'their duty to bring order and control to Ireland', to make the land productive.[91] Places that had previously been named for wolves were now transformed into pastoral paradises with English names to reflect the new status quo, wolves both real and cartographical erased from landscape and map alike.[92] As one nobleman wrote in 1633, an area of Cork that had been 'a mere waste bog and wood serving as a retreat and harbour to woodkernes, rebels, thieves and wolves' had been transformed by the early seventeenth century into 'as civil a plantation as most in England'.[93]

Spurred on by their distaste for wolves and wilderness, efforts to remove wolves by the English invaders became more and more concerted. In the late 1500s the Lord Deputy of Ireland was presented with a proposal to tackle the 'infestation', which suggested that trapping and killing should be required of land tenants as part of their leases.[94] But more pressing issues apparently got in the way. The Nine Years' War of 1593 to 1603 – the result of an Irish rebellion – put the wolf problem on the backburner. The English eventually emerged victorious, though they were the renewed rulers of a country 'almost entirely laid waste and destroyed', where 'terrible want and famine oppressed all'.[95] The wolves, meanwhile, were considered more problematic than ever before, with records describing them growing in number and becoming bold while the country lay in ruins,[96] sallying forth from their woodland and mountainous retreats to 'attack and t[ear] to pieces' those who were 'weak from want'.[97] Hence, in 1611 another 'act for killing wolves and other vermin' was proposed,[98] and rewards were offered for the delivery of wolf corpses. The bounty system attracted hunters from England,[99] one of whom was given a generous £3 per wolf head (around £650 today) in 1614, and was granted licence to employ four men and multiple wolf-hunting dogs in every county of Ireland for seven years, meaning he could have had a total of 128 men and 768 dogs at his disposal at any one time.[100]

But the efforts were largely in vain. By the middle of the seventeenth century, there were perhaps still as many as 500 to 1,000 wolves still roaming Irish shores.[101] Numbers had likely increased when, following the Irish Rebellion of 1641, the human population sharply decreased due to the war, plague, and famine,[102] diverting attention away from the wolves, creating plenty of ready-made corpses for them to feast upon, and providing habitat in areas where people had disappeared.[103] By the early 1650s, 'ravening wolves' were apparently even eating orphaned children, and were commonly seen venturing into the outskirts of Dublin.[104] In response, efforts to eradicate wolves began in earnest during Oliver Cromwell's dictatorship of 1649 to 1653. Cromwell banned the export of Irish wolfhounds in 1652,[105] because they were too valuable for hunting the wolves that 'do much increase and destroy many cattle in several parts of this Dominion'.[106] The following year the Cromwellian government issued a 'declaration touching wolves', which outlined methods 'for the better destroying of wolves, which of late years have much increased in most parts of this nation'. According to this declaration, the 'commanders in chief and commissioners of the Revenue' in 'several precincts' of Ireland were required to 'consider of, use and execute all good ways and means, how the wolves [...] may be taken and destroyed'.[107]

Under this scheme, extraordinary bounties of £6 for a female wolf, £5 for a male, 40 shillings for a hunting juvenile, and 10 shillings for a young pup were offered; around £1,182, £985, £394, and £99 respectively today.[108] By 1656, a total of £3,847 and 5 shillings, around £777,146 in today's money, had been spent on the wolf extermination campaign,[109] a figure possibly including bounty payments as well as the equipment required to facilitate wolf hunting and trapping.[110] A year later, a Member of Parliament for Wicklow spoke of bounty payments of as much as £10 for female wolves (£1,916 today), describing them, along with native rebels (upon whom there were also bounties), as 'beasts to destroy, that lay burdens upon us'.[111] As in Norman England, territory in Ireland was also parcelled out on the condition that its occupiers – both farming tenants and the elite who were granted land – killed wolves.[112] A man named Edward Piers, for example, held a lease for state-owned land in Meath on the condition that he kept numerous dogs for wolf-hunting and employed people to utilise them, who were to hunt at least three times a month and who were to kill no less than fourteen wolves (as well as sixty foxes) within five years.[113]

But the wolves continued to cling on. Reports of their depredations continued, with an officer in the English army describing in 1657 how wolves continued to 'exceedingly infest' Antrim and 'haunt and undo the tenants', with 'keepers and [...] gun men' still being forced 'to watch the wolves [...] almost every night' some eight years later.[114] The early 1660s also saw the government petitioned to encourage wolf hunting, with one man apparently even seeking permission to utilise a unique – though unfortunately unspecified – new method in 1663.[115] Even as late as 1673, the Anglo-Irish naturalist Robert Boyle (1627–1691) observed that hunters could 'make a gain, if not an entire livelihood, by killing of wolves' in Ireland.[116]

And so the wolf cull went on, and people continued to make money from bloodied corpses.

Eventually, these protracted efforts proved successful. By the year 1700 wolves were much rarer than they had been a century before,[117] the survivors forced into increasingly marginal areas to find almost non-existent shelter from a far-reaching and hostile human presence.[118] In these disconnected pockets of habitat, the divided remnants of the wolf population eked out their existence until, eventually, they could sustain themselves on a genetic level no longer.[119]

It is around this point that the narrative shifts from 'infestations' and 'great stores' of wolves to comments on their scarcity and absence. A document written in 1683 notes that although wolves were once 'very numerous' in Leitrim, they had since become 'very scarce' thanks to bounties paid from a 'hearth tax' (whereby money was collected for every hearth in each parish member's home – in this case, a tuppence).[120] A letter written in 1698 by a Cork alderman even notes that wolves had been transformed from 'noxious and hateful' to 'game and diversion',[121] presumably due to their rarity. They also apparently no longer posed a significant threat to sheep – a 1669 tract written by Irish friar Anthony Bruodin (d.1680) observes that despite not being completely absent, wolves were no longer a source of concern for sheep farmers in Co. Clare.[122]

Like their Scottish neighbours, the last wolves of Ireland likely stayed as far away from people as they could within the limited space available to them.[123] And yet, just as in Scotland, England, and Wales, sensational stories of the 'last wolf' in Ireland (or in a particular locale) poured forth from all over the country.[124]

One, from Co. Mayo, depicts the 'last wolf' apparently having the audacity to enter a bothy, where it nearly extinguished a man's fire by shaking itself dry before promptly attacking him as he lay in bed. The man managed to trap the wolf in a blanket before killing it, though the bothy burnt down during the commotion. It is telling of the reliability of this tale that other versions of this story do not feature a wolf at all, but a 'wild cat'.[125] Other tales see the 'last wolves' attacking domestic animals – even a horse is credited with killing the last wolf in Tyrone after it set upon her foal.[126]

This latter tale contradicts another story of the 'last two wolves' in Tyrone at an unspecified date. These wolves were on the hitlist of a professional wolf hunter after killing some sheep. According to the story, the hunter took a 'little boy' with him to wait for the wolves at the sheepfold where they had previously committed their ravages. Reasoning that the wolves would enter via the gaps in the wall at either end of the sheepfold, the hunter leaves the boy to guard one end while he departs for the other. But, in a well-worn cliché, the boy promptly falls asleep after being told not to do exactly that. Right on cue, one of the wolves appears. Luckily, a wolfhound accompanying the boy has stayed alert, and leaps at the approaching wolf with a great roar. The boy promptly awakes and heroically kills the wolf with a spear to the neck. The hunter then returns, weighed down by the head of the second wolf.[127]

An equally implausible story surrounds a carving from Ardnaglass Castle in Sligo. This relief sculpture, which portrays a dog killing a wolf, was presented to the Royal Irish Academy in 1841, the donor claiming that it commemorated the death of the last wolf in Ireland. In response to its depredations on local livestock, the wolf was supposedly chased mercilessly by a local chieftain and his famous wolfhound, before finally being killed in a stand of pine trees at the foot of the Tireragh Mountains. This chieftain then supposedly commissioned the carving to commemorate his great victory, and the place where the wolf was killed was named Carrownamadhoo, meaning 'dog's quarter'. The relief is in fact medieval in date.[128]

It is frequently said that the last wolf in Ireland was killed in 1786. This individual was allegedly hunted down by a man named John Watson after it attacked sheep on Mount Leinster.[129] Supposed evidence to corroborate this legend comes from a portrait of Watson's son, who sits with his feet on what is said to be a wolfskin.[130] But the date at which this story is set is implausibly late, and the tale appears to be derived from family tradition repeated to investigators of the legend by a member of the Watson family.[131] Perhaps the story was told to impress children and grandchildren, who sat upon the skin as they listened to chilling tales of their forefather's heroic exploits. That is, if the pelt belonged to a wolf at all – the rug in the portrait does not look particularly lupine.

In all likelihood, wolves had all but disappeared from Ireland by the turn of the eighteenth century, though small numbers may have survived in remote areas a little beyond 1700.[132] The actual 'last wolf' who ever walked the shores of Ireland, just like the last wolf in Scotland, probably 'lived secretly and silently on the peripheries of human life and died of old age after failing to find a mate' – alone, unnoticed, and unmourned.[133]

The nineteenth-century antiquarians who penned many such 'last wolf' tales invented or distorted the facts, apparently feeling 'that they acquired a degree of kudos in discussing the subject', and so 'v[ied] with each other to produce the latest record'.[134] Even by the second year of the twentieth century authors desperate to say something new were still upping the ante, with one speculating that 'it is just barely possible, indeed, that an isolated specimen or two of the breed may yet exist among the pathless wilds of Connemara, or some equally savage district'.[135] As recently as the 2010s it has been speculated whether small wolves could be found on Achill, a large island off Ireland's west coast, until the early twentieth century.[136] These theories are based on a comment by British explorer Henry Johnston (1858–1927), who described seeing dogs on Achill 'exactly resembling the wolf in colour, in brush, in the shape of the ears, and in the arrangement of the masses of hair along the line of the back'.[137] This comment was seized on by cryptozoologists who thought that these dogs could actually be wolves affected by island dwarfism.[138]

Old myths of savage wolves living in savage wildernesses are not easily extinguished.

The 'Uninfested' South

For those living in the newly 'uninfested' lands of England and Wales, the wolf had become 'other'. It remained a fearful creature who haunted the dark wilderness at the edge of civilisation, but for the English and Welsh of the early modern period, that dark wilderness now lay beyond seas and borders or was firmly consigned to the mists of time. The world had been brought firmly under their own dominion, and they were no longer haunted by the fear – irrational though it may have been – of being stalked and killed by a ferocious wild beast.

They also no longer had to worry about wolves eating their livestock. In the wolf's absence England and Wales became prime sheep-farming territory, which was now a much easier vocation since the animals did not need to be so closely shepherded, nor housed overnight for protection.[139] The 'great flocks' and the 'great revenues made every yeare' from these woolly moneymakers were safe. According to English writer Edward Topsell's (c.1572–1625) *History of Four-footed Beasts*, the economic worth of sheep even increased because of the absence of wolves. Topsell notes that English wool was so prized and 'commended [. . .] highly' because its sheep's coats were 'soft and curled', since 'they are neither annoyed with the fear of any venomous beast, nor yet troubled with Wolves'. Thus, the 'peaceable quiet wherein they live[d]', a landscape so perfect that they 'quench[ed] their thirst' with the very 'dew from heaven', '[did] breed in them the better wool'.[140]

Buoyed by their success in establishing themselves firmly as the masters of this wolfless landscape, efforts to reinforce human control over the natural world continued at an ever-greater pace. Industrialisation, exploitation, management, and commodification were seen on a grander scale than ever before. The need to feed, house, heat, and meet the demands of a continually increasing population transformed the landscape. The scientific advances of the period also facilitated a greater understanding of the natural world, increasing the ability to bend it to human will and need.[141]

Any animals that hindered human progress, no matter how small, were almost as intolerable as wolves and were treated in kind. As seventeenth- to eighteenth-century English clergyman Edmund Hickeringill (1631–1708) commented, 'so noisome and offensive are some animals to human kind, that it concerns all mankind to get quit of the annoyance'.[142] Such 'offensive' animals, upon which bounties were set during the sixteenth century, included a vast array of birds and mammals, from foxes, rats, ravens, weasels, and moles, to otters, ospreys, and even hedgehogs and kingfishers.[143]

As in Topsell's account of the wonderful wool of England's sheep, tones of quiet pride in the tamed landscape run throughout early modern descriptions of the wolfless land. William Camden, for example, praised the Peak District's 'grassy hills and vales', which sustained 'many flocks of sheep' because 'there is no more danger now from wolves which in times past were hurtful and noisome to this country'.[144]

Rumours even emerged that wolves had disappeared from England because its landscapes were entirely unsuitable for the species altogether, reflecting a belief that wolves did not and had never belonged. During a diplomatic mission to Prague in 1577, esteemed English poet Philip Sidney (1554–1586) was asked by German scholar Philipp Camerarius 'whether it was true [...] that England cannot endure wolves, either bred in the country, or brought thither out of other places', because of 'some hidden property and natural antipathy' of the country. Sidney, however, attributed the wolf's absence to the 'wisdom of our kings', whose extermination campaigns meant that England had been 'clean rid of' wolves for 'a long time'.

Sidney was apparently under the impression that wolves had been absent for much longer than was the case, describing how an unspecified king 'a great while since' had ordained that banished criminals could pay for their crimes with wolf heads and tongues, a system which ultimately led to the wolf's demise in England.[145] As we saw in Chapter 3, it seems that this legend was inspired by medieval stories about the tribute of wolfskins demanded of the Welsh by the Saxon king Edgar.[146] *Holinshed's Chronicles* also ascribed the absence of wolves in England and Wales to this monarch, and addressed the thorny issue of where the wolves that lived in England in the following centuries had come from by suggesting that they were foreign introductions.[147]

Sidney thus explained to Camerarius that wolves were perfectly capable of surviving in the English climate and were doing so even at that very moment, being kept 'in parks of great lords, who send for them out of Ireland and other places, to make a show of them as of some rare beast'.[148] Wolves had been transformed from malign enemy to exotic curiosity, becoming firm fixtures of both menageries and the collections of travelling exhibitors. As Harrison noted in *Holinshed's Chronicles*, wolves were brought back to England 'from beyond the seas for greediness of gain, and to make money only by the gazing and gaping of our people upon them, who covet oft to see them, being strange beasts in their eyes, and seldom known [...] in England'.[149]

Even the English royalty appeared to be fascinated these 'strange beasts' from times gone by. In 1592, a German visitor to the Tower of London menagerie described seeing, along with lions which he claimed were around a century old, a 'lean, ugly wolf, which is the only one in England; on this account it is kept by the Queen'.[150] Whether or not this wolf was in fact 'the only one in England' at this time, it was soon one of many which were imported from all over the world to be sold by enterprising merchants, including 'a large wolf from Barbary', 'a pair of wolves from Maryland', and 'a white wolf from Greenland' imported by a merchant named Michael Bland in 1726 and 1727.[151]

These animals ended up in private collections, serving as living symbols of their owners' power and wealth.[152] One such collector was the second Duke of Richmond (1701–1750) who, in his expansive gardens in West Sussex, housed five wolves along with two tigers, a lion, a civet cat, two Greenland dogs, four bears, several monkeys, three racoons, and even an armadillo, among other species.[153] Wolves also seem to have been kept at Chirk Castle in Wrexham in the late 1600s and early 1700s.[154]

Other animals were taken on the road to delight and terrify visitors to these mobile menageries. Gilbert Pidcock, a travelling showman, toured around England with 'two stupendous ostriches ... a Bengal tiger, a young Lioness, a real laughing Hyena, a ravenous Wolf, an African ram [...] and the double-jointed Irish Dwarf'.[155] The wolf's 'ravenousness' was apparently now the subject of interest rather than fear, so long as it was safely caged. Another animal merchant named Joshua Brookes (who in 1767 had an impressive catalogue of animals for sale including an ostrich, lions, a porcupine, a camel, leopards, and some wolves from Saxony),[156] even became interested in crossbreeding wolves with dogs in partnership with an anatomist and menagerie owner named John Hunter (1728–1793). The results of these experiments were sold to aristocrats, one of whom bought a puppy born from a union between a female Pomeranian and a wolf, an animal that looked like a wolf and was skilled at killing deer. The success of these experiments led Hunter to correctly surmise that wolves were the ancestors of dogs.[157]

As wolves grew ever more exotic, their fur became more and more popular. Several hundred wolf skins were shipped from Ireland to Bristol annually throughout the sixteenth century,[158] and their value rose exponentially. While they were worth 1½ pence each (around £14 today) in 1492, by the mid-1500s they fetched more than five times as much, at 8 pence each. A bumper import of 961 pelts which arrived in

Bristol between 1558 and 1559 was therefore worth just over £32 – the equivalent of more than £15,000 today.[159]

Although it was still less expensive and less highly prized than the striking fur of the lynx and leopard, which was used to line the clothing of royalty and nobility, wolf fur was often used in the lining of men's clothing.[160] An inventory of the belongings of the Earl of Pembroke drawn up in 1561 mentions numerous garments edged with a variety of animal furs such as red squirrel, sable, rabbit, and wolf, including black satin gowns and a russet silk coat edged with grey and black wolf fur.[161] Wolves had been transformed from a local scourge to an exotic species, found only in distant lands over the sea and in the far north.

But for all their interest now that they were deemed exotic animals, or perhaps precisely *because* they had been transformed from native to exotic, wolves were considered to belong firmly behind bars or far away in wild foreign lands. The inhabitants of predator-free Britain would not tolerate wolves returning to their countryside. As Philip Sidney noted, both natural and human-made protections were purportedly in place to prevent their incursions into the south from Scotland, including deep rivers and 'mighty garrisons' defended by 'great store of dogs'.[162] For those who imported wolves, meanwhile, Camerarius noted that it was 'forbidden upon grievous penalties to let them escape out of their enclosure', and 'very sharply forbidden to bring or to fetch wolves from any other countries, that might store England again with the vermin of which it had been delivered'.[163]

Yet rumours also swirled that wolves had not gone extinct at all. Swiss academic Guy Miège (1644–c.1718) noted that he had been 'credibly informed' that in some parts of England, 'some wolves from time to time have been discovered'. Such incidents were a rare occurrence, but supposedly inspired the local populace to 'rise' together 'as it were against a common Enemy', to hunt down and kill the offending animal.[164] In reality, this was likely little more than 'wolf hysteria whipped up against feral dogs or animals imported for hunting'.[165] A French naturalist, apparently on the authority of an English doctor, also describes a creature supposedly present in England called the 'sea wolf', noting that 'as much as the English have no wolves on their land, nature has supplied them with a beast on the banks of their sea [...] strongly resembling our wolf'. This animal apparently had a 'build, fur, head [...] and tail closely approaching the terrestrial wolf', and 'if it were not that it throws itself rather on fish than on [... sheep], one would say it was just like our very predatory beast'.[166]

Sea-wolves and hysteria aside, with real wolves more or less gone, people turned their attention to ridding England of 'wolves' in human form. As well as criminals, who were described by one contemporary of William Shakespeare (1564–1616) as 'serpents or wolves, or worse than both',[167] usurers were often compared to wolves. In a tract written in 1572, English diplomat Thomas Wilson (1524–1581) stereotypically praises the efforts of King Edgar to rid England and Wales of wolves, which 'was a good deed surely, and a gracious proclamation'. Wilson then goes on to suggest that removing moneylenders from England would be 'a greater good deed to this land, than ever was done by killing of wolves', because usurers are 'greedy [...] wolves in deed that raven up both beast and man, who while they walk in sheepskins, do covertly devour the flock of England'. He ultimately argues that just as wolves were slaughtered during Edgar's reign, so too should usurers 'suffer the pains of death, or be banished this realm for ever'.[168] Shakespeare drew upon the same imagery in his portrayal of Jewish usurer Shylock in *The Merchant of Venice*, in which Shylock demands a pound of flesh from a debtor who has failed to repay him. The usurer is described accordingly as 'wolvish, bloody, starved, and ravenous', the soul of a wolf that had been 'hanged for human slaughter' having 'Infused itself' in Shylock while he was still in utero.[169]

Catholics were also often portrayed as wolves by Protestants amid the religious upheavals of the Reformation, becoming a 'metaphor for the perceived Popish threat'.[170] Catholic priests in particular were frequently the victims of this metaphor, with one graphic image on a mid-sixteenth-century pamphlet depicting a wolf-headed Stephen Gardiner, Bishop of Winchester (c.1483–1555), who was involved in the persecution of Protestant Reformers. This wolf-priest is seen tearing into the throat of a strung-up sacrificial lamb, a symbol of the innocent Protestants. The lamb's blood spurts over a wolf-headed congregation of Catholic clerics wearing lamb skins, an evocation of the 'wolf in sheep's clothing', who hold their cups in gory anticipation. More trussed-up sheep are seen awaiting the same fate below the wolf-priest, labelled with the names of executed members of the Church of England, while a gleeful-looking devil surveys the scene. This grisly reimagining of the Eucharist (the sharing of bread and wine to represent Christ's body and blood) evokes the violence and greed of the Catholic clergy. The role of these 'wolves in sheep's clothing' was to protect and shepherd. But in their pursuit of power they slaughtered innocent lambs, the true followers of Christ, and led their own followers to damnation.[171] For the Protestants, these Catholic 'wolves' belonged in England no more than the real wolves that had been extirpated.[172]

Even cancer was known as 'the wolf' because, as Topsell wrote, 'it consumes and eats up the flesh in the body next the sore, and must every day be fed with fresh meat [...] or else it consumes all the flesh of the body, leaving not so much as the skin to cover the bones'.[173] Though this euphemistic lupine name had been in use since the 1200s, it became ubiquitous four centuries later. Some even thought that cancerous tumours took the form of a literal wolf which inhabited and fed upon a person's body, growing in size and strength as the afflicted wasted away. Such beliefs led to the use of an unusual treatment in which raw meat was placed next to the cancer to draw the 'wolf' out, enticing it to eat the meat rather than the person. One woman who underwent this treatment reported seeing the wolf 'peep[ing] out' of the cancerous ulcer, 'gaping to receive' the meat.[174] This metaphor was also used in non-medical contexts, in which the cancer-wolf symbolised characteristics such as covetousness and ambition. These bestial instincts 'consumed' a person at the expense of their human sense of morality, just as the cancerous 'wolf' ate a person from the inside out, hidden destroyers of human flesh and spirit alike.[175]

It was not just the cancer-wolf that could devour a person. Stories of people being killed and eaten by men transformed into wolves reached English shores in pamphlets detailing the werewolf trials of Europe. Although the fear of witches from which accusations of werewolfery stemmed was also rife in England, the English witchcraft trials were entirely devoid of lycanthropes, perhaps since belief in and stories about werewolves tended to be associated with places where wolves were (still) found.[176] But this did not make the stories any less appealing. Morbid curiosity fanned the flames and fervour surrounding these tales of foreign magicians who could metamorphose into wolves.

The best-known pamphlet that circulated in England at this time was titled *A True Discourse Concerning the Damnable Life and Death of One Stubbe Peeter* (1590). It describes (and illustrates, in gory detail, for the illiterate) the exploits of an infamous 'werewolf' from Cologne named Peter Stumpp. According to the pamphlet, Stumpp made a pact with the devil, from whom he received a magical girdle which would allow him to transform into a 'greedy devouring wolf [...] with most sharp and cruel teeth, a huge body, and mighty paws' at will. Since werewolves at this time were thought to be humans in true wolf form (as opposed to the half-man half-wolf that the werewolf is often depicted as today), in his transformed body Stumpp could 'work his malice on men, women, and children' 'at his pleasure', especially since he also killed lambs and other livestock, so that it seemed that a real wolf was on the rampage. He went on to commit the 'most heinous [...] murders' of people who had wronged him in the past, 'pluck[ing] out their throats and tear[ing] their joints asunder'.[177]

Stumpp acquired such a taste for murder that he was soon killing with abandon, gaining 'such pleasure and delight in shedding of blood, that he would night and day walk the fields, and work extreme cruelties' on any man, woman, or child unfortunate enough to cross his path. All the while, he appeared as a person of 'comely habit' when in his human form, even as he scoped out his next victim. Over the course of twenty-five years he killed and consumed countless people including his own son, brain and all, as well as two pregnant women, 'tearing the children out of their wombs' and 'eat[ing] their hearts panting hot and raw, which he accounted dainty morsels and best agreeing to his appetite'. Finally, after too many years spent discovering stray arms and legs left behind from Stumpp's ravages, the people of Cologne apprehended the malefactor. Stumpp confessed to his crimes and was sentenced to die by being placed on a wheel, where his skin was burned with red-hot pincers, his flesh was pulled from the bone, his limbs were broken, and he was finally beheaded, along with his mistress and daughter. The pamphlet ends with a list of four names of 'witnesses that this is true', which could also be attested by 'diverse others that have seen the same'.[178]

Despite this assurance that Stubbs' metamorphosis was real, there was much debate as to whether the transformation of people into wolves was possible at all, or whether people who believed they could turn into wolves were simply deluded. Even James VI and I (r. as King James VI of Scotland 1567–1625 and as King James I of England and Ireland 1603–25) waded into the debate, writing in his 1597 treatise on magic, the *Daemonologie*, that the belief that one could turn into a wolf, and the wolfish behaviour of those who thought as much, 'proceeded but of a natural superabundance of melancholy'.[179]

Physicians such as Robert Bayfield (fl.1668) were likewise agreed that lycanthropy constituted a medical rather than a spiritual issue – as Bayfield wrote, 'wolf-madness, is a disease, in which men run barking and howling about graves and fields in the night [...] and will not be persuaded but that they are wolves'. The disease was said to have a physical effect, causing 'hollow eyes' and 'scabbed legs and thighs' that were 'very dry and pale', but episodes could be resolved by bloodletting and the administration of a potion.[180] But stories such as that contained in the pamphlet about Peter Stumpp undermined the stance of the Church, physicians, and kings alike that true transformation was not possible. It is unknown whether the general populace at this time believed in werewolves, but the popularity of pamphlets about the European werewolf trials suggests at least a fascination with – if not popular belief in – these creatures in early modern England.[181]

While lynx and bears more or less faded into cultural obscurity and apathy when they disappeared, myths about wolves lived on even after they had vanished. The fear of wolves as savage, bloodthirsty creatures perpetuated by myths and legends lasted far longer than the animal itself, metamorphosing the wolf in southern Britain from a creature of flesh-and-blood to one of paper-and-ink. That perceptions of them did not change upon their extinction implies that anxieties and fears about wolves were less about the animal itself and more about its cultural life – the meanings imbued in it by people and the language of hatred and fear in which its being was encoded, which already ran too deep to be excised. For early modern people, wolves were 'wild', 'fierce', 'bold', 'greedy', 'degenerate', 'ungentle', 'unhonest', 'untameful', 'harmful', 'furious', 'insatiable', 'mad', 'cruel', 'terrible', and 'inhuman'. They were 'flesh-eater[s]', 'blood-lovers', 'glutton[s]', 'teeth-gnasher[s]', 'treacherer[s]', 'robber[s]', 'ravener[s]', 'snatcher[s]', and 'blood-sucker[s]'.[182] While Topsell described these words as 'most clear demonstrations of [the wolf's] disposition',[183] this list speaks far more to the one-sided war which continued to be waged against wolves with violent pen and vicious ink, even long after they had lost the battle to survive.

Wolves in the New World

When British colonists first encountered wolves in the Americas, it must have been like entering a primeval nightmare. The presence of the very beasts that they had spent so long attempting to exorcise from their homelands was such a surprise and concern to Scottish cartographer John Ogilby (1600–1676) that he could not fathom how they got there, commenting in 1671 that only 'domestic and other creatures fit for human use and sustenance' were transported to the Americas, leading him to wonder who exactly 'would load their ships with lions, tigers, bears, wolves, foxes, and other serpents and voracious beasts?'. He refuted that 'the angels conveyed them thither', because he could see no reason why God would not 'plant men there in like manner', since 'the earth [was] created for human use'.[184]

The view that wolves did not belong in the New World was widespread. The newly settled continent was considered a 'feared and fearful expanse in which all sorts of dangers lurked and as a place that had to be subdued and dominated if [the colonists] were ever to bring civilization to their new homeland', as anthropologist Garry Marvin puts it.[185] Wild and savage beasts were intruders to be exterminated and rightfully replaced by livestock.[186]

Wolves directly threatened the viability, security, and ultimately the survival of the early colonies. With their introduced domestic livestock, the settlers had inadvertently provided these predators with a far easier meal than the wild animals they were used to preying upon.[187] The colonists in turn had no idea how to protect their animals against the attacks of wolves and were often too busy tending to crops to do so, creating the perfect storm for depredation.[188] Compounding the problem was a huge decrease in bison numbers due to human hunting, changes in the ecosystem, and disease,[189] leaving the wolves little choice but to prey on the cattle that took their wild prey's place.[190]

Wolves undoubtedly made a tough existence even more difficult. The gravity of the issue is made clear by the numerous records from the early New England colonies describing the loss of livestock to wolves, which take up considerable space in town records. As one chronicler writing of New England in the early years of colonisation noted, 'among the trials of the first settlers, there were none more irritating than the destruction of sheep and swine by the wolves and bears', with 'whole flocks of sheep' often 'slaughtered in the night'.[191] Another colonist writing in 1634 similarly described wolves as 'evil, and of most annoyance to the inhabitants' of the New World; 'the greatest inconveniency the country has'.[192] For some, the situation became dire – in 1643, wolves had killed off all but twenty sheep belonging to the New Amsterdam colony on Manhattan Island.[193]

The emotional toll was enormous. The colonists' survival was staked upon their livestock, into which they had poured precious time and money. To awake in the morning to find these animals killed during the night – perhaps many of them in one fell swoop – would have been a tough pill to swallow. Wolves were thus quickly cast as verminous thieves. The loss of livestock was a fact of life duly accepted when it was due to disease, inclement weather, and other natural causes, but wolf attacks were personal.[194]

Just as when they preyed upon livestock, wolves were also considered thieves who stole the rightful property of humans when they killed deer. Numbers of native ungulate species had declined rapidly as a result of hunting by the colonists to supplement their diet, leading local governments to introduce hunting seasons by the end of the 1600s.[195] Yet despite humans causing the severe declines in moose, deer, and elk numbers, the blame was placed upon wolves. One colonist wrote that it simply did not bear considering 'what great multitudes [of deer] would increase, were it not for the common devourer the wolf'.[196]

Though some colonists had journeyed from Scotland, Ireland, and other European countries that still had local (though heavily diminished) wolf populations, the majority were English. Very few settlers, therefore, had any direct experience of living alongside wolves. What they did have, however, were stories and ideas of these rapacious and evil beasts.[197]

Some settlers were surprised to learn that wolves were not so deadly as they had feared or expected. In a letter penned in 1638, one colonist wrote that 'our greatest enemies are our wolves, but yet [they] flee man' and are 'afraid of us',[198] while in a description of New England from 1634, writer William Wood (c.1580–1639) noted that the wolves of America were 'in some respect different from them of other countries; it was never known yet that a wolf ever set upon a man or woman'.[199] One settler even successfully shooed a wolf pack away from a deer carcass and stole it for himself.[200] Of course, these wolves were no different from the European and extinct British wolves who also attacked people only rarely, but English settlers were so used to hearing stories of man-eating beasts that it was a complete surprise to encounter wolves that did not set upon humans at every opportunity.

But for the colonists, who lived on a spiritual diet of Christian scripture, wolves did not only pose a physical threat. Just as Christ sent his disciples 'forth as lambs among wolves', the colonists too thought of themselves as sheep in Christ's flock, embarking on a sacred mission to civilise an untamed wilderness full of wolves.[201] Spiritual and physical wolves were confused. The real animals were treated as though they were the devil incarnate, 'creatures of the godless wilderness that the colonists believed they had a moral duty to subdue', emblems of an untamed wild and the ultimate 'natural enemies to civilization'.[202]

Though wolves were apparently 'so numerous' that there was felt to be 'little hope of their utter destruction',[203] as one colonist miserably commented, that did not stop the settlers from trying. Extensive bounties were put in place, just as they had been in the Old World. Wolf habitat was destroyed, hunts were frequently undertaken, professional hunters were hired, and some colonists were even required to keep dogs to 'destroy vermin such as wolves'.[204] Efforts to kill wolves were so concerted that some communities spent as much on bounties as they did on everything else combined. This provided 'yet another reason to hate wolves', who were 'a continuous drain on community resources that governments felt compelled to continue' lest the population rebound and their efforts were for naught.[205]

Despite the 'little hope' that wolves would disappear, the effect of these measures was astronomical. Wolves had little experience with humans who found their presence unwelcome, having lived for hundreds of years alongside a Native American population who did not see wolves as enemies but as respected cohabitors of a landscape that belonged to all species. Native Americans did not keep livestock, instead hunting for meat alongside the native predators. They killed wolves only infrequently, for their pelts and for medicinal purposes, rather than because of interspecies conflict.[206] But the newcomers were vastly different. They severely depleted the wolf's prey, carved out land from wolf territories and claimed it for themselves, drove wolves away from their newly established settlements, and perhaps even caused packs to starve under the pressure.[207] Most of all, they hated wolves and they killed them for it. As the English colonists pushed further westwards, wolf numbers slowly but surely declined.[208] By the close of the 1700s, less than a century after the colonists had arrived, wolves had all but disappeared from New England.[209]

It was not only wolves that were vehemently persecuted as unwelcome intruders on the 'civilised' New World which the colonists were attempting to build. Native Americans who, for the most part, revered, respected, and sometimes identified with the wolf,[210] were frequently compared to or equated with wolves by the colonists. Like wolves, they were deemed wild and savage, incompatible with and threatening to the rapidly expanding 'civilisation', even 'heretical to the emerging working landscape'.[211] To the settlers they were essentially wolves in human form who, if they fought against colonial rule or stole livestock, were characterised as verminous predators preying upon the rightful custodians of the land.[212]

There was about as much love lost on the Native Americans as on wolves. In the minds of the colonists, the displacement and murder of the native human population was as justified as the eradication of wolves,[213] and they, too, were hunted with impunity. As one New England clergyman wrote in 1703, in an attempt to justify hunting Native Americans with dogs, 'they act like wolves and are to be dealt withal as wolves'.[214] Like the wolf population, numbers of Native Americans were soon rapidly dwindling in the face of sustained persecution.

Those who survived were conscripted into the mission of lupicide. Colonists in Virginia managed to kill two birds (or wolves) with one stone by offering Native Americans one cow per eight wolves they killed, which not only reduced the wolf population with little effort by the colonists themselves, but was also regarded as 'a step to civilizing' the Native Americans and 'making them Christians'.[215] Some colonies even demanded that Native Americans paid a tribute of wolf heads as 'an expression of their submission to colonial rule', even at the same time as Native Americans themselves were being decapitated.[216]

Over the following centuries, a campaign of destruction and death was enacted across continental America, until wolves were all but wiped out from the lower 48 states. They were reviled as a stain on the land which deserved nothing less than to be wiped clean from its surface. Attitudes towards them are perhaps best summed up by Theodore Roosevelt's (1858–1919) famous description of the wolf as 'the archetype of ravin, the beast of waste and desolation'.[217] The only good thing about them was that they could be hunted for sport, an activity so thrilling that, so Roosevelt said, 'nothing more exciting [...] can possibly be imagined'.[218]

By the 1850s, traps were produced on an industrial scale. They were used not only for catching animals whose fur could be sold, but to capture predators.[219] Countless wolves were caught in such traps and, despite often being 'cowering and submissive', were subject to unimaginable torture. The lucky ones were shot or beaten to death. The less fortunate were tied up and showed off to a cheerful community who revelled in the capture of a 'wanted criminal'; set upon by dogs; pulled along behind horses until they were exhausted, bloodied, and broken; torn apart when tied to two horses running in opposite directions; or burned alive.[220]

They may have all but disappeared from British and Irish shores, but wolves had not escaped from the fear and hatred that had festered and driven them to their deaths in the Old World.

CHAPTER 6

A Post-wolf World

Something is missing from the forest. The mountains and moors are silent. The wind sweeps through the heather, lonely, howling. Its call goes unanswered.

Of once-warm bodies, now only cold tracks remain. These old pawprints, baked into clay long ago, bespeak absence. Hollowness. Rain, sand, and seawater fill the holes.

Caves in craggy hillsides lie empty, carpeted by the crunching, bleached-white and sepia-yellow bones of the dead.

Life continues, but in a strange and strangled way. The deer are bold, uninhibited except by the roar of engines or the pop of a gun, nibbling saplings down to nubs. Plastic protective tubing falls away, useless, turning to litter in its failure. Fewer and fewer winged insects fill the summer air with their droning. The precious few wildcats that creep through the forests of the north lose their homes to logging and their lives to loneliness. Field after field after field of sheep after sheep after sheep create a white polka-dotted patchwork. Monocultures of grass and plastic deserts of astroturf carpet the gardens of suburbia. The air is poisoned by smog vomited from the stone cigars of industry. Rivers carve out new courses in the absence of their orange-toothed, wood-eating engineers, bursting their banks and spilling over where they are not welcome. Somehow, nature itself is to blame; humans are the victims.

Wolf Land has lost its wolves, and its way.

Wolfless Islands

When Wolf Land became wolfless, both the natural world and the human relationship with it were fundamentally and irrevocably changed.

While the grazing of ruminant species (first aurochs, wild cattle, and boar, which were outlived by red and roe deer) had maintained a mosaic landscape for thousands of years, wolves helped to keep the numbers of these nibblers in check. In so doing they eased grazing pressure on woodlands, keeping it in balance with scrubland and open wood pasture.

Deer were in turn replaced by livestock. In Scotland, the popula tion of domestic species grew exponentially following the Highland Clearances of the eighteenth and nineteenth centuries. This brutal and merciless series of enclosures saw families unceremoniously and sometimes violently removed from land they had occupied for generations. Tens of thousands of people were evicted from hundreds of thousands of acres by the landowners, who wanted to create empires of livestock.

The overwhelming numbers of cows, goats, and particularly sheep that replaced the human population put the already struggling forests under ever-greater pressure. Under such conditions, it was impossible for the dwindling tree numbers to recover.[1] Some have suggested that the enclosures could not have happened if wolves had not been eradicated – it would make little business sense to fill the landscape to the brim with such easy pickings for a hungry predator.[2] But without wolves and the fear of losing their animals to lupine depredations, landowners were free to increase their flocks and herds exponentially, to allow them to roam far and wide without the need to pay shepherds to care for them.

Later, when the value of wool decreased, sport shooting became fashionable, and the coming of the railways allowed people to make faster and easier trips to the north for their leisure, estate owners turned to grouse shooting and deer stalking to make money. The sheep were in turn cleared out to make way for deer forests, 3.5 million acres of which had been established in the Highlands by 1914.[3] Deer numbers increased, replacing sheep as the grazers preventing woodland from regenerating. The late 1800s also saw the creation of commercial plantations, where native tree species were ripped out and replaced with 'monotonous plantations of black fir'.[4]

Though the Highland Clearances were perhaps the most infamously cruel acts of enclosure ever to happen in Britain, they were not an isolated occurrence. Thousands of tenants were also forced out of their homes in the Scottish Lowlands, while the early years of the nineteenth century saw some 200,000 acres of Welsh land enclosed by Acts of Parliament.[5] Landowners in Ireland likewise enacted widespread clearances during and after the Great Famine (1845–52), evicting tens of thousands of families in less than a decade so that their land could be consolidated and converted to pasture.[6]

In England, the process of parcelling out land to create pasture was already well under way by the time of the Highland Clearances. By the early sixteenth century, sheep in England had 'destroy[ed] and despoil[ed] fields, houses, [and] towns' thanks to landowners who 'enclose[d] everything as pasture; […] thousands of acres within one hedge', as Sir Thomas More (1478–1535) described in his *Utopia*.[7] Though it is estimated that only 2% of England was enclosed between 1500 and 1600, over 25% more was to follow between 1600 and 1760, bringing the total proportion of enclosed land to 75% of England's area.[8] While not all enclosed land will have been used for pasture,[9] the exponential increases in livestock numbers indicate that much of it was – by the early 1800s, England housed almost 2.5 million cows, between 12 and 14.5 million sheep, and almost 2 million pigs.[10]

At the same time, wealthy landowners sought to shape the natural world not only for utilitarian purposes, but for aesthetic pleasure as well. Just as land was enclosed for farming, so too did wealthy and influential men cast aside entire villages so that they could claim the ground they occupied for parkland.[11] These landscapes became landscap*ed* – idyllic natural scenes carefully designed to look like a painting, dotted with the hallmarks of human presence to indicate that this was controlled, cultivated, and tamed land which had been bent firmly to human will. Immaculate lawns were interspersed with artificial watercourses and carefully positioned trees, shrubs and flowers. And, of course, they were dotted with sheep.[12] As landscape designer Humphry Repton (1752–1818) put it, gardens were spaces 'appropriated to the use and pleasure of man', 'cultivated and enriched by art' yet 'made to assimilate with […] the landscape of nature'.[13] Wolves were not welcome here.

So too were the 'wilds' of the Scottish Highlands romanticised by artists and writers, who transformed the vast, wild expanse imagined in previous years into a magnificent – and distinctly wolfless – landscape, a perception which conveniently ignored the fact that much of the Highlands were already cultivated and largely treeless. This was in no small part a product of the perceived 'civilising' of the Highland landscape and its people following the failed Jacobite rebellion in the mid-1700s, when laws were passed to suppress Highland culture.

But even as the landscape was increasingly disrupted by intense grazing pressure and pruned into painting-worthy submission, rural life was ever-more romanticised by an ever-more urbanised and industrialised population. The unnaturalness came to seem natural. The image of the carefully pruned, 'chocolate box' countryside – even the semantics were adapted to accommodate for these human-created, cultural landscapes – has been glorified ever since as the 'natural' state of the world. As then-Conservative Party leader Stanley Baldwin (1867–1947) said in a 1924 speech, the stereotypical scene of a green and pleasant countryside punctuated by 'the sight of a plough team coming over the brow of a hill' was thought of as a 'sight that has been seen in England since England was a land', a scene that 'strike[s] down into the very depths of our nature, and touch[es] chords that go back to the beginning of time and the human race'.[14] Never mind that the British landscape looked rather different at 'the beginning of time' and had done for most of its existence since, until it was reshaped almost beyond recognition through human intervention of the last few hundred years. The myth of the timeless rural idyll persists just as strongly today, even in an age where 12% of 18- to 24-year-olds in Britain have never seen a cow in person.[15] Today, as since the birth of the 'chocolate-box' countryside, wolves are perceived to have no place in this vision of idealised British rural life. As one Colin Matheson put it in 1943, even 'the most ardent exponents of fauna protection […] will see in the reduction of the wolves only a necessary step in man's control of his environment'.[16]

Outlaw Wolves

The only places where wolves belonged in this newly tamed landscape were zoos and travelling shows, firmly behind bars. Wolves were brought from all over the world to entertain – in a rather horrified sort of way – the inhabitants of wolfless Britain and Ireland. The knowledge that such savage and rapacious creatures had once stalked their shores evoked fear and relief in equal measure for those who paid to see them. Seeing these 'vicious' beasts provided tangible evidence that their ancestors had been undeniably wise to wipe out such a dangerous predator.

This impression was heightened by the poor state in which the wolves were kept. They were often mistreated and rarely fed so that they could be more easily manipulated,[17] ensuring that they did not disappoint customers who wanted to see ravenous and savage animals that lived up to their reputation. It certainly worked – as one observer of a 'tame wolf' act at Westminster Aquarium in 1887 noted, when he 'looked at these fine but terrible creatures', he 'could almost realise what it might be to be followed by a pack of hungry wolves, and how fearful the dread of hydrophobia [the fear of water – a symptom of rabies] from such like bites must be to all dwellers on their native [...] lands'.[18]

Given the sensationalism of these exhibits, it is little wonder that hysteria ensued when wolves attacked zoogoers from behind bars. Although attacks by other menagerie species were common, such incidents were framed rather differently than attacks by wolves. If elephants or lions injured people, their actions were put down to an ill-tempered individual or blamed on unfortunate provocation by the victim. But when wolves attacked, it was interpreted as evidence of the species' intrinsic savagery and maliciousness.[19] Even their manner of eating was thought to betray their ravenous, uncivilised, and insatiable natures, especially when compared to animals such as a chimpanzee at London Zoo named Sally, who had been trained to eat 'with a spoon, like a little lady'.[20]

Wolf escapes were likewise covered in the news with a rather hysterical edge. In 1834, when an 'enraged' wolf escaped from the Tower of London menagerie and chased a dog into a nearby apartment, a woman and her children bolted in 'great terror', leaving the wolf to seize the dog by the throat before 'sucking the blood out of his prey', according to a report in *The Standard*.[21] In 1888, eight wolves escaped from a circus in London, going on to attack a horse 'in a most ferocious manner' before 'ravenously devouring' the carcass,[22] sparking panic that they might graduate to attacking people.[23] Though it was an admittedly more horrifying incident when a wolf attacked a man in the zoo at Dublin's Phoenix Park in 1858, seizing his hand in its mouth, newspaper accounts rendered the wolf in sensationally vampiric terms: the 'warm blood' which was 'dropping into the brute's mouth' was said to 'render its thirst for blood more eager'.[24]

Even when they merely trotted off to enjoy their newfound freedom, escaped wolves were reliably treated as though they were mass-murderers who had broken out from Broadmoor. In 1843, a black wolf imported from Russia escaped from a menagerie in Coventry, which 'excited no small commotion and terror' amongst the visitors and caused a stampede for the exit.[25] Three years later, another errant wolf that did little more than kill a cat after escaping in Camden was described as a 'dreaded' animal which 'prowl[ed] about the neighbourhood' and had made its 'lair' in a nearby field, causing 'some considerable alarm' among local residents. The 'ferocious creature' was soon shot before being bludgeoned to death.[26] An escaped wolf also caused 'a good deal of consternation' as it roamed nearby Peckham the following year, despite merely wandering around the area without doing much of anything. Though this wolf was 'old and much diseased', the idea of peacefully capturing the 'intruder' was seemingly unthinkable. The poor animal was instead pursued and eventually killed by 'a body of men armed with guns and sticks'.[27]

Of the numerous wolf escapes that occurred throughout the nineteenth and twentieth centuries, perhaps most famous is the animal known as the Hexham or Allendale wolf, which ran rampant in Northumberland in the final months of 1904. Over a three-week period the wolf killed multiple sheep, and panic bubbled as parents worried that the wolf might attack their children. The 'Hexham Wolf Committee' was established in response, and a well-known pack of foxhounds, a famous bloodhound, a Hungarian wolf hunter, and a 'big game hunter' from India were all put on the case. Yet all efforts to capture and kill the beast proved fruitless even when, harking back hundreds of years to the policies of the premodern era, a £5 bounty was placed on its head.[28]

The reward was never claimed. Instead, just before the year's end, a dead wolf was found on the train tracks near Carlisle. Northumberland breathed a sigh of relief. Locals were so relieved, in fact, that a graphic photograph taken by a railway porter of the wolf's bloodied and battered body circulated widely on postcards, a souvenir of the terrifying yet thrilling incident. The wolf's identity was, and remains, a matter of some confusion. At first, it appeared to be a pet wolf pup belonging to one Captain Bains, which had escaped three months prior. But the captain insisted that the animal killed by the train was not his pet, which he said was much smaller and younger than the corpse he was presented with. Though it would perhaps appear that

Captain Bains simply wanted to avoid paying compensation for all the sheep that the wolf had killed, sightings of a rogue wolf in the area continued over the next few weeks, until interest in the case eventually waned.

Even in more recent years, despite far more stringent controls on who can keep animals and which animals they are allowed to have,[29] there are still not infrequent reported sightings of wolves and wolf-like creatures throughout Britain and Ireland. The press seizes on such incidents – after all, they make for a good headline. In 2017, *The Sun* wrote about a man who heard a wolf howling in the wee hours near his hotel in the Outer Hebrides. It mattered little that experts ascribed the howling to a wolf–dog hybrid, which were known to be bred nearby. The man who heard the creature, and many online, believed it to be a genuine wolf howl.[30]

But some sightings are very real. Recent years have seen multiple escapes from zoos and wildlife parks, whether deliberately caused by animal rights activists or because of accidental damage to enclosures. As in the nineteenth century, the response to these escapes is often highly disproportionate to the threat that these animals actually pose. When a wolf named Torak escaped from his enclosure at the UK Wolf Conservation Trust in 2018, armed police were soon deployed, one of whom guarded the entrance of a local primary school on lupine lockdown. Caution was clearly sensible, but a 12-year-old (in lupine terms, very elderly), tame wolf perhaps did not warrant such extremes. As one expert from the University of Warsaw noted at the time, 'it is the wolf which is in danger, not that the wolf is dangerous' – Torak was at risk of being attacked by domestic dogs, expiring from extreme hunger, and being hit by a car. Perhaps most threatening were the unwitting residents of Wolfless Land whose peace he had intruded upon, whose 'lost connection with large carnivores' infused the escape with a whiff of hysteria,[31] disproportionate fear, and an itchy trigger finger. Wolves, after all, are far less dangerous statistically than the millions of dogs owned in Britain and Ireland, who bite or injure an average of 250,000 people and kill several each year,[32] along with tens of thousands of sheep. In contrast, the overwhelming majority of wolf–human encounters in the wild see the animal fleeing. Thankfully for Torak, his sojourn into freedom ended in him being peaceably put on a lead and returned to his enclosure, none the worse off for his adventure.[33] Most other escaped wolves are not so lucky.[34]

Beasts from the East

If the people of eighteenth-century Britain and Ireland were in any doubt that eradicating wolves had been a wise decision, reports of an enormous man-eating wolf terrorising the French province of Gévaudan surely swayed them. From 1764 to 1767, several hundred people (mostly women and children) were attacked across the region by what seemed to be an enormous animal. Many were killed and eaten. Gory accounts of the beast's ravages – including one particularly gruesome attack in which the creature strangled a woman before biting her neck, sucking the blood from her corpse, and taking her head away with it – soon sparked widespread panic.

The identity of the monstrous beast was – and remains – a mystery. Initial reports described it as having a strange mixture of features, including rust-coloured fur with a black stripe along its back; a three-foot high stature; a long tail; and variously the snout of a pig, the muzzle of a calf, or the head of a cat. Some thought it the 'monstrous product of a cross between a bear and a female wolf', or an animal of leonine parentage crossed with an indeterminate species.[35] As the months turned into years, the identity of the beast continued to be steeped in speculation and scaremongering, with claims surrounding its characteristics becoming wilder and wilder. By 1765, one newspaper stretched all credulity by describing it as possessing a lion's mouth, a boar's tusks, and the scaled back of a fish.[36] Clearly, such a chimera did not exist. But *something* was hunting, killing, and eating the people of Gévaudan, evading capture despite the efforts of thousands of hunters, including a company of dragoons of the French army.

One of the prime suspects was a wolf, and so in 1765 specialist wolf hunters were dispatched by the Crown to track the animal down. They too failed, and so King Louis XV's own gun-bearer, François Antoine, was sent to take up the mission. After months of searching Antoine finally felled a wolf in September of 1765, which was so large that he initially thought it was a donkey.[37] He declared this animal to be the infamous beast, subsequently verifying his claim on the authority of several survivors of its attacks, but conveniently setting aside the fact that the surgeon who autopsied the animal found no evidence that the creature had eaten any people.[38] Nonetheless, the wolf's mate and pups were also tracked down and killed for good measure.[39]

But by December the attacks began anew, and it soon seemed that le Bête de Gévaudan – or, perhaps, one of multiple Beasts of Gévaudan – was still at large. All renewed attempts to hunt and kill the creature were unsuccessful (though almost 100 wolves were killed in the process[40]) until, in the summer of 1767, a hunter named Jean Chastel felled a second beast. A post-mortem report described this creature as 'quite different in shape and proportion from the wolves normally seen in this country' (as attested by 'several hunters and many knowledgeable people'), noting a strange mix of characteristics including a 'monstrous' head; a strange membrane in its eyeballs; a mixed pelage of red, grey, and white which 'the hunters had never seen on wolves' before; and a strangely malleable skeleton.[41] Whatever this animal was – and theories as to the identity of the beast(s) include wolf, dog, wolf-dog, hyaena or, assuming the second animal killed was also not 'the' Beast, a human serial killer[42] – the death of this second creature finally put an end to the three-year long nightmare that the Gévaudan had endured. All the while, English newspapers had regularly reported on the beast's reign of terror, no doubt leaving their readers utterly relieved that they had long since eradicated their wolves.

Even such stories did not stop some from attempting to bring the wolf back, however – in a manner of speaking. Finding that fox hunting was not to his satisfaction, a keen eighteenth- to nineteenth-century sportsman named Colonel Thornton (1751/2–1823) is said to have had ambitions to import wolves from abroad to set free on the Yorkshire Wolds.[43] Other hunters of the newly dubbed 'British Wolf', the fox, had different schemes.[44] A newspaper report from 1810 suggested that a fox hunt had attempted to crossbreed wolves with foxes, 'in order to improve his ferocity and add such strength as would effectually deter any shepherd [...] from attacking him single handed', a new quarry which could be reserved for their exclusive hunting pleasure.[45] Luckily for the residents of Wolfless Land, such schemes were fated to fail – if, in fact, they did take place at all.

Fiction becomes Fact

It was not just wolves in (and out of) menageries, zoos, and newspaper reports that left an impression upon the inhabitants of wolfless Britain and Ireland. By the Victorian period, the history of the wolf in Britain and Ireland itself had started to take on the mythologising overtones that have dogged the animal's reputation throughout time.

None so damaged the reputation of the British and Irish wolf nor so distorted its history as much as James Harting (1841–1928), a natural historian who dedicated over a third of his 1880 book *British Animals Extinct within Historic Times* to the animal. Harting filled his tome with a mixture of fact, fiction, and everything in between. His 90 pages on the wolf were so inflated largely because of the abundance of myths and legends that he unquestioningly repeated. Such accounts make good stories on a winter's night gathered around the fire, important cultural artefacts which deserve to be preserved as such, but they are far from the facts Harting presents them as.

Yet Harting has long been considered an authority on the subject of Britain and Ireland's lupine history. Fiction became fact in the telling, retelling, and endless repetition of his diatribe. Even Harting's quite clearly prejudiced wording – including such phrasing as 'infest' and 'rabid droves' – is often repeated verbatim today, preserving a language that frames wolves as diseased swarms, spreading through and infecting the landscape like a plague of insects. Though wolves in Britain and Ireland may have succumbed to rabies at times, they were far from forming 'rabid droves'. Nor is it a sign of Harting's scientific objectivity that the only alternative meaning of 'rabid' that he could have intended was 'furious, raging; wildly aggressive or violent'.[46]

But Harting is perhaps not to blame so much as the zeitgeist of his time. His 'distortion of the reality of wolves',[47] as Jim Crumley puts it, was common among the Victorians, who unquestioningly accepted the anti-wolf sentiments perpetuated both by their forefathers and by one another under the guise of scientific inquiry. We have already seen how Harting owed some of his fictional 'facts' to the false authority of the Sobieski-Stuarts, and much of the rest of his work has equally questionable sources. Harting was the product of a time when wolves were unquestionably accepted as 'rabid droves' that had once 'infested' the landscape, and stories which fit into this portrait were rife and ripe for collection, never mind whether they were genuine or not. In a culture of lupophobia, such fictions could easily appear to be fact.[48]

The constant repetition of Harting's work and his treatment as an authority on the subject of British and Irish wolves only serves to legitimise such stereotypes and loaded language. Today, wolves are the object of falsehoods perpetuated by those who dislike them and oppose the idea of their return to Britain and Ireland. They play into stereotypes about the wolf's savage nature and recycle the same rhetoric of violence and destruction that has punctuated discourse about the animal for centuries, their way paved by the 'authority' of Harting and those who repeat him.

The modern media is often guilty of perpetuating these falsehoods.[49] Headlines frequently describe wolves as 'killer beast(s)', 'ruthless killers', 'bloodthirsty', or 'at war' with people, and anthropomorphise these animals by attributing decidedly human characteristics to them, such as ruthlessness or warmongering. Such rhetoric is often used in the context of the wolf's possible return to Britain and Ireland. A 2023 article in *The Spectator* titled 'Reintroducing Wolves to Britain is Pure Insanity', for example, asked 'should we release packs of ravenous wolves into the English countryside?' in its opening sentence.[50] This article plays into age-old stereotypes of the wolf's insatiable hunger and savagery, as well as ideological notions that the natural state of Britain's landscapes is an idyllic 'countryside' in which wolves have no place. The answer to the question is rhetorical: a resounding 'no'. Or, as the article's author put it, 'of course not, are you insane?'.

Metaphorical uses of the terms 'wolf' and 'wolves' also reflect poorly on the animal, despite having nothing to do with the animal itself. Recent news stories have seen gang rapists described as 'ravenous pack[s] of wolves' and the perpetrators of anti-social behaviour likened to 'wild bunch[es] of wolves',[51] while terrorists who act alone are commonly figured as 'lone wolves'.[52] As with the headlines about real wolves, these news stories capitalise on the wolf's reputation as a rapacious and insatiable beast.[53] They zoomorphise the perpetrators to distance them from humanity, rejecting and disowning behaviour deemed uncivilised, savage, and predatory. Wolves serve as the repositories of such behaviour, onto whom everything we fear about our own species and what we are capable of is projected,[54] despite the fact that wolves do not rape, shoot air rifles, throw rocks at cars, or commit terrorist attacks, as did all the 'wolfish' people in these reports.

Wolves are also the victim of Western political agendas of nationhood, migration, and xenophobia. Immigrants are likened to 'invading' wolves and are framed as parasitic 'swarms' and 'infesting' 'invaders',[55] in much the same way as Harting described wolves themselves as an 'infestation' of 'rabid droves'. Fears of the wolf returning to old haunts to 'claim' land and food that should, for many, belong to people,[56] have become intwined with fears of savage and uncivilised 'migrants coming in from the wild', 'invaders' who steal land and resources seen as rightfully belonging to the natives.[57]

Right-wing groups in Europe claim that both wolves and immigrants (especially those from Islamic countries) are 'foreign' and do not belong. They 'blam[e] the EU for its open borders to wolves and migrants', expressing a desire to outlaw 'burqas and wolves' and to hunt down and eradicate migrants and wolves alike. In so doing, they dehumanise unwanted migrants to justify their violent removal, while simultaneously anthropomorphising wolves as deliberate invaders and criminals.[58] It is likely that such rhetoric would surround any potential return of the wolf to Britain, where the reintroduction of 'foreign' wolves from Europe would undoubtedly be one focus for opponents, just as the Canadian wolves translocated to Yellowstone National Park in 1995 were branded 'foreign' by those who opposed their reintroduction.[59]

Yet paradoxically, wolves are also used as a metaphor for those who would resist the 'invasion' of unwanted immigrants, who see themselves as wolves defending their territory against the incursions of trespassers. In 2016, Donald Trump's supporters warned immigrants that 'the wolves are coming … and you are the hunted'.[60] Others have styled themselves as wolves in 'a politics of excessive nationalism', as academic Peter Arnds puts it, with the wolf functioning 'as a positive symbol standing in for strength, courage, aggressive invasion of new territories, [and] pack organization'.[61]

Perhaps most infamous is the adoption of wolf symbolism by the Nazis. During the Second World War, German submarines were called 'wolf packs' by both Germans and the Allied forces, Hitler referred to the SS as 'my pack of wolves', and he styled several of his headquarters in lupine terms, such as the *Wolfsschanze* ('Wolf's Lair') in East Prussia. Hitler was also particularly keen on adopting a lupine identity himself. His first name came from the Old German Athalwolf ('noble wolf'), and he signed letters with and referred to himself as 'Herr Wolf' or 'Conductor Wolf'. Even a decade before he was in power he had already begun styling himself as a lupine saviour of the German people, writing in a 1922 newspaper article that 'a wolf has been born, destined to burst upon the herd of seducers and deceivers of the people'.[62]

Such rhetoric has made its way to Britain, with the creation of a Neo-Nazi white supremacist group called the White Wolves. In 1999, around the same time as the racially motivated London bombings that this organisation claimed credit for, the White Wolves also sent death threats which warned 'all non-whites' to 'permanently leave the British Isles before the year is out', claiming that any 'Jews and non-whites who remain after 1999 has ended will be exterminated'. The notes ended by stating: 'When the clocks strike midnight on the 31st of December 1999 the White Wolves will begin to howl and when the Wolves begin to howl the Wolves begin to hunt.'[63] Thankfully, nothing came of this threat.

The Reign of the Cultural Wolf

Though many people in Britain and Ireland today may fear wolves, what they in fact fear is not the animal itself but an imagined, ghostly echo of the real-life creature.

This is in no small part because we in Wolfless Land experience (and, one might say, obsess over) wolves in so many different ways: in films, television shows, documentaries, newspapers, novels, non-fiction, children's fairy tales, even on t-shirts and countless other products. All, in fact, but the most important: in the flesh. Studies have shown that the less people have to do with real wolves, the more they fear and mythologise them. In the absence of experience, stories are all that we have left to go on. And when it comes to the wolf, there are plenty of those. No other animal has lingered so long in our collective imagination. Few people continued to tell stories about beavers or lynx after they disappeared. Bears were reimagined as cuddly, honey-eating co-conspirators of children. The wolf was less lucky.

Stories of ravenous, duplicitous wolves are the bread and butter of children's tales. One of the best known, the story of Little Red Riding Hood, has frightened and delighted children all over Europe since the late seventeenth century. The oldest known printed version of this tale is found in a 1697 collection of fairy stories by French author Charles Perrault, although it is likely that it has folkloric roots which stretch back much further. Perrault's tale was translated into English not long after its initial publication, and its popularity and ubiquity were further reassured by its inclusion in the well-known fairy tales of the Brothers Grimm.

The details of this early French story are much the same as in the versions we know today. As Le Petit Chaperon Rouge travels through the forest to visit her sick grandmother she is intercepted by the wolf, who easily extracts the naïve little girl's destination from her. Sly Mr Wolf gets there first and tricks the grandmother into letting him in before eating her at record speed, putting on her clothes, and getting ready for his second course. Little Red arrives and gets into bed with the wolf, and they exchange the infamous 'what big ears/eyes/teeth you have' dialogue. The wolf, exclaiming that his teeth are "all the better to eat you with!", swiftly gobbles her up. The end. Except for the moral, which Perrault makes explicit:

> Young children, as this tale will show,
> And mainly pretty girls with charm,
> Do wrong and often come to harm
> In letting those they do not know
> Stay talking to them when they meet.
> And if they don't do as they ought,
> It's no surprise that some are caught
> By wolves who take them off to eat.

Such 'wolves', Perrault suggests, may not initially appear to be predators. As well as the obviously dangerous, sharp-toothed wolves, there are some who are 'not the savage kind, / Not howling, ravening or raging'. Instead, these wolves appear to be 'friendly' and 'engaging' and are 'softly-spoken and discreet', but they follow vulnerable 'young ladies' home, and are 'the most dangerous wolves of all'.[64]

The Grimm brothers' version, first published in German in 1812 and translated into English a little over a decade later, features a different ending which sees the wolf brought to justice. After eating Little Red Cap and her grandmother whole, the satiated wolf slumps back into bed to take a nap. A passing huntsman hears his thunderous snoring and decides to make sure the old lady to whom the house belongs is alright. But instead of granny, he discovers the 'old sinner' who he has 'been looking for […] for a long time'. Realising that the wolf has eaten the old woman, the huntsman takes a pair of scissors and cuts open its stomach. Out jump a frightened but relieved Little Red, followed by her grandmother, all while the wolf somehow remains asleep. Little Red then fills the wolf's body with stones, which kills the wolf when he wakes up and tries to run away. Unlike in the French version, the Brothers Grimm leave their audience to work out the moral of the story for themselves.[65]

Though the Grimm version of the tale leaves the wolf's metaphorical identity to one side, Perrault makes his wolf not only a would-be man-eater but a potential rapist and child groomer to boot. An early oral version of the story even featured a scene in which the wolf tells Little Red to throw her clothes into the fire and get into bed with him naked.[66] The real wolf has clearly been left far, far behind, replaced by an animal who embodies the worst 'appetites' of humans. The fear of wolves that is evident in and perpetuated by the story of Little Red Riding Hood is, as nature writer Barry Lopez so memorably put it, little more than human 'self-hatred' and 'anxiety over the human loss of inhibitions that are common to other animals who do not rape, murder, and pillage' – ultimately, 'fear of one's own nature' which is displaced onto the lupine 'scapegoat'.[67] But it is wolves themselves that suffer most from these metaphors, not the predatory humans that they have come to symbolise. It is wolves who end up being the objects of fear, detested and reviled, their bellies stuffed full of stones and their bodies pumped full of lead.

Thanks to lupine fairy tales like Little Red Riding Hood, wolf howling has become a trope in children's media to create a frightful atmosphere – an animalistic version of 'it was a dark and stormy night'.[68] The animals themselves also often make an appearance to add a sense of danger, fear, and drama, particularly in Disney films. The deep dark woods in *Frozen* (2013) and *Beauty and the Beast* (1991 and 2017) are home to wolves that leap from the shadows to attack the protagonists, the yellowed whites of their wide eyes glinting madly, their jaws slavering, their noses wrinkled as they snarl and snap at their potential meal.

For most children in Britain and Ireland, the wolves of fairy tales and films are the first they encounter. From the earliest age they, like children all over the world, learn that wolves are evil man-eaters ruled by their cavernous stomachs, bloodthirsty and savage tricksters who prey upon innocent children and animals, attacking anything or anyone foolish enough to stray into the deep dark woods where they lurk. Anti-wolf promotional material from places where wolves survive only affirms these fears. Children in Washington were undoubtedly terrified when, in 2015, an organisation known as the Washington

Residents Against Wolves erected billboards featuring images of deer, a cow, a dog, and a smiling child below a blown-up pair of menacing lupine eyes nestled in jet-black fur. The text reads: 'The Wolf... Who's Next on Their Menu?'[69]

Conversely, children do not often learn that wolves are an important part of healthy ecosystems, and at worst may think that they are damaging to the natural world because of their predatorial behaviour. One study of children's attitudes towards wolves in Slovakia and Turkey found that over 50% of children did not understand why wolves were ecologically important, and 20% actively believed that they were not. Almost 60% thought that nothing bad would happen if wolves became extinct.[70] This study found that such attitudes were attributable to negative depictions of wolves in children's stories, which led to increased fear and decreased sympathy. Watching nature documentaries, however, was associated with decreased fear.[71]

Yet even documentaries do not always present an impartial portrait of wolves. Although such programmes give insight into the true nature of the much-maligned animals, footage of wolves is often portrayed in such a way as to emphasise their role as predators, to play into stereotypes of their ravenousness and savagery, or to add a thrill of danger in situations where there was little to none. A documentary series that reveals rather more about its human subjects than its wolves, *Mr and Mrs Wolf* (2009), capitalised on lupine clichés to create drama and fear. In this programme, self-proclaimed 'wolf man' Shaun Ellis attempted to introduce his girlfriend to the captive pack of wolves into which he had integrated himself (and of which he had allegedly become 'alpha'). The narrator emphasises that Ellis 'joined the pack [...] to prove that wolves are more than just teeth and stereotype'.[72] But any possibility of this programme rewriting the narrative is immediately dashed by overlaid snarling sound effects and transitional shots comprising closeups of bared wolf teeth, all stained with a blood-red filter.

A less sensationalised documentary was Gordon Buchanan's 2014 series *Snow Wolf Family and Me*, which saw the filmmaker spending time observing and interacting with a family of Arctic wolves that had never before encountered people. Though largely presenting a balanced portrait of these extraordinary animals, footage of a close encounter between man and wolf is stereotypically framed to send a shiver of dread down the viewer's spine. Gordon lies on the ground, propped on his elbows, to 'build a whole new level of trust' with the wolves, as he puts it. But as the background music darkens from twinkly and upbeat to tense dissonance, he wonders whether he's 'taken this too far'. Yet, as the camera pans from the back of Gordon's head to his feet and the musical tension mounts, we see not a snarling, predatory beast, but simply a curious wolf in a tentative, cautious posture, rather understandably sniffing the boot of the odd bipedal stranger who has chosen to prostrate himself just metres from the pack. The wolf swiftly backs off, looking visibly nervous. On the BBC Earth YouTube channel, the clip is sensationally titled 'Dangerously Close to Wolves', despite the situation being about as dangerous as it would be to lie down next to any other large, wild animal – and far less so than many other species.

The fear of wolves is clearly not left behind as we exit childhood, nor if we foray into the world of supposedly objective documentaries. Likewise, in the books, films, and even video games we consume as adults, we continue to encounter wolves which hark back to the fairytale beasts of our childhood bedtime stories.

Epitomising and perpetuating the use of wolves as a source of horror and dread is Bram Stoker's 1897 novel *Dracula*. As Stoker's protagonist, Englishman Jonathan Harker, travels closer to Dracula's home in the Transylvanian wilderness, the wolves start to close in. They are unseen at first, but they make themselves known by their persistent and terrifying howling until a ray of moonlight finally illuminates the pack's 'white teeth and lolling red tongues', a sight of pure 'horror' which paralyses Harker with fear. Harker's increasingly wolf-haunted approach to Dracula's castle is an early sign of the infamous vampire's close relationship with these animals. Upon his arrival at the castle, Harker learns of the vampire's appreciation of the 'music' of these 'children of the night' and his ability to 'enter into the feelings of the hunter' as others cannot – all of which is a hint of his true, predatorial nature.[73] We later discover that Dracula himself can transform into a monstrous wolf and, in a curious but oft-overlooked plotline, he develops a relationship with a wolf at London Zoo, prompting this previously peaceable creature to break out and follow the 'call of the wild'.

While in many cases wolves simply feature as a monstrous cliché of the horror genre, stereotypical portrayals of lupine ravenousness and savagery are sometimes used in contexts which frame these traits as the wolf's natural behaviour. One of the most notorious examples is the film *The Grey*, which was released in 2011 to vociferous protestation from pro-wolf groups. Set in the freezing Alaskan wilderness, *The Grey* follows a group of men who survive a plane crash, only to be picked off one by one by a vicious pack of wolves. The film capitalises on conventional portrayals of wolves as bloodthirsty mankillers – the sounds of snarling and howling and the images of glinting teeth and ethereally glowing eyes peering from the darkness building tension and horror throughout the film.

Though they look lupine, these animals are almost unrecognisable as wolves in their behaviour. Rather than fleeing from the presence of the human interlopers and the fires they light, the wolves approach the group of men and attack them unprovoked. The 'alpha' wolf (a stereotypically large, black beast) 'sends' an 'omega' wolf to 'test' the men, as though wolves can draw up and communicate battle plans with one another. The pack are described as 'man-eaters' which 'don't give a shit about berries and shrubs', despite an obvious plot hole: where did the pack pick up a taste for human meat in the middle of the Alaskan wilderness? Planes surely don't crash that often. The film attempts to negate this question by attributing the wolves' predatoriness to the presence of the men close to their den: "I thought wolves were scared of people," one man says – "Not if we're near their den. They're not scared of anything then," Liam Neeson's character replies. Here, the film plays into the wolf's well-known territoriality to gloss the outlandish behaviour of these filmic animals as natural and, by extension, believable.

In response to criticisms of the film's portrayal of the animal, the distributor uploaded a factsheet about wolves to their website. But despite this, and despite the absurdity of some of the behaviours and motivations attributed to the wolves in the film, a casual viewer from a culture of lupophobia could all too readily believe the misinformation which drives its plot. Wolves attacking hapless survivors in the icy wilderness of the far north is itself a cultural trope ubiquitous enough that it could easily be accepted as true.

Following a similar storyline is *The Long Dark* (2014), a video game in which the player assumes the role of a crash-landed pilot attempting to survive in the freezing Canadian wild. If exposure, lack of potable water, or starvation do not kill them, aggressive wild bears and wolves certainly will. The hapless player who strays too close to these animals is swiftly set upon, and unless they are successful in wrestling the attacker off, they will soon bite the dust. As in *The Grey*, this game plays into stereotypical portrayals of wolves as bloodthirsty and violent creatures who attack on sight, rulers of the wilderness who will not tolerate the incursions of man. Just as the distributor of *The Grey* attempted to allay criticisms of its depiction of wolves, *The Long Dark* opens with a disclaimer that they 'have taken liberties with the portrayal of wildlife behaviour', since they 'are not attempting to create "realistic" wildlife behaviour in the game'. They assure the player that they 'know that wolves do not typically attack people'.

But wolves attacking people is an unfortunately common cliché in video games. In countless instances, especially in games with fantasy or (pseudo-)medieval settings such as *The Witcher* (2007–15) and *Elden Ring* (2022), wolves are a frequent nuisance to players as they explore the map, despite the worlds of such games being populated by far stranger and stronger fantastical beasts. In the hugely popular game *Fortnite*, meanwhile, wolves with scarlet eyes lurk around woodland waiting to attack the player, who must kill or be killed. When the *Fortnite* update that introduced these deadly wolves was rolled out in 2021, the International Wolf Center expressed concerns that, just as 'for hundreds of years, myths and fairy tales have misinformed the world's children about wolves', *Fortnite* too was 'spreading […] myths' about the dangerousness of these animals and 'glorifying violence against wolves to a whole new audience' of young people.[74]

As the IWC noted, 'mischaracterizations like these can have a profound effect on people's perceptions about wolves'.[75] Even if we consciously reject the notion that wolves are bloodthirsty killers, and rationalise that such stereotypical portrayals of wolves in culture have little to do with the wolf of the real world, it is often difficult to separate the biological wolf from the cultural wolf – particularly when we have far more experience of the snarling, slavering latter. Clearly, it will take a lot to overturn the wolf's reputation for boundless malice and rapacity. Venturing deep within the thickets of hatred and fear, crushed under the weight of misrepresentations, and enveloped by the long shadow cast by Mr Wolf, it is difficult to imagine that the biological wolf's image will ever be rehabilitated.

A Changing Narrative

But despite a reign of terror spanning thousands of years, the Big Bad Wolf is now beginning to enter his redemption arc. Authors, filmmakers, and audiences alike are starting to question whether he really is so big and bad after all.

Perhaps the first person to probe this question was Rudyard Kipling who, in his 1894 collection of children's stories *The Jungle Book*, rethought the wolf's reputation and inimical relationship with humankind. Animated by Disney in 1967, *The Jungle Book* follows a young child named Mowgli who is raised by a pack of wolves in the Indian jungle. Mowgli has a happy existence as a wolf, learning to survive in the wild from his adoptive parents and littermates. Although Mowgli befriends other predatory animals during his time in the jungle, including much-loved Baloo the bear and long-suffering Bagheera the black panther, that this human child is protected by the wolves against the antagonist tiger, Shere Khan, marks a significant departure from traditional perceptions and portrayals of wolves.

The wolves in *The Jungle Book* are represented not only as familial, nurturing, benevolent, and accepting towards an outsider from across the species barrier, but also as animals 'liv[ing] in a supportive, orderly and disciplined society'.[76] Considering the wolf's reputation for bloodthirst and child-eating at this time, as is only too evident from Harting's *British Animals*, the cultural impact of *The Jungle Book* upon the wolf's image was profound.

Inspired by the novel, a new branch of the Scouts called the Wolf Cub scheme was established in 1916. The Wolf Cubs were organised according to the pack dynamics portrayed in *The Jungle Book*, with the adult leader(s) of the 'cubs' known as the Akela, the name of the patriarch of the wolf family Mowgli is raised by. As with Mowgli and his adopted wolf siblings, the Cub Scouts were expected to respect and obey their Akela, to work as a team and to be loyal to their 'pack', and to be disciplined, sentiments enshrined in the 'Wolf Cub Law' which dictated that 'The Cub gives in to the Old Wolf. The Cub does not give in to himself'. Both Kipling and Robert Baden-Powell (1857–1941), the British Army officer who established the Wolf Cubs, saw past the bloodthirsty wolf that loomed so large in the culture of their time. They found positive and even admirable traits in these maligned animals, new ideas about wolves which they subsequently passed on to thousands of young boys.

Kipling's reassessment of the wolf was inspired by the strange stories emerging from India of children 'raised' by wolves in a manner akin to Romulus and Remus. Numerous instances of this phenomenon were reported in British newspapers which, despite doubts having since arisen as to their truth, were 'universally believed' at the time, according to Kipling's own father.[77] Though stories of children raised by wolves have captivated people for thousands of years, they were perhaps even more alluring for the inhabitants of post-wolf Britain and Ireland, whose only experience of these creatures was in lore and literature which consistently framed the animals as savage wild beasts, decisively dangerous to humankind. Tales of wolves suppressing their (apparently) fierce natures by nurturing human children were wondrous and fascinating precisely because of the wolf's reputation for everything that these benign creatures were not.

Many children's authors have since taken up Kipling's mantle. Numerous books, from Katherine Rundell's *The Wolf Wilder* and Laurence Anholt's *Eco-Wolf and the Three Pigs* to Jean Craighead George's trilogy *Julie of the Wolves*, feature people living harmoniously alongside wolves; wolves as or becoming heroes rather than villains; wolves as benign or vital inhabitants of fantasy worlds; or just wolves being wolves.[78]

As Lucy Jones observes in her book *Foxes Unearthed*, there is power in narratives like these. Like his lupine cousin, the fox was traditionally represented and understood in negative ways throughout history, having a reputation for craftiness, cunning, slyness, predatoriness, and thievery, and often serving 'as a symbol for many of society's ills'.[79] But all this changed when a new fox came to town, with the publication of Roald Dahl's 1970 children's book *Fantastic Mr Fox*. Dahl did not change the fox's cunning nor his predatoriness, instead reframing these previously reviled traits as virtues. The fox becomes the hero, his crafty exploits to steal a feast's-worth of chickens from several unpleasant farmers celebrated as ingenuity and 'fantasticness', rather than being condemned as predatory thieving.

Yet despite the prevalence of 'good wolf' stories, it is difficult to find tales about wolves in which their predatoriness is celebrated in the same manner as in *Fantastic Mr Fox*. Stories of 'little good wolves' who do not want to be bad like all the rest are far more common.

Dahl's story accepted and celebrated the fox's natural traits even in the face of pushback from his editors, who suggested that Mr Fox should steal eggs rather than killing and eating chickens. Mr Fox is 'a hero in spite of' – or perhaps even because of – 'his natural carnivorous behaviour',[80] not because he overcomes it. As controversial as foxes still are in Britain and Ireland, perhaps this is what the wolf needs too.

Though their natural predatory behaviour has rarely been reappropriated in a positive way, much new writing about wolves, particularly in non-fictional contexts, has seen their character reassessed. These narratives draw us away from the Big Bad Wolf of our childhoods and encourage us to move towards a future that is far kinder to wolves, both on a cultural level and in the real world. The wolf has become a powerful symbol of wildness and freedom in many narratives, particularly for groups who have historically been maligned. Irish writers have explored the island's loss of wolves as concurrent with and equivalent to attacks on their national identity,[81] both at the hands of English invaders in the seventeenth century – the colonists not only wiped out wolves but placed restrictions and bans on the use of the Irish language.[82] These authors seek to reclaim the English colonisers' denigrating identification of the Irish as wolves and of Ireland as a wolf-infested wasteland.

Feminist writers have also metaphorised women as wolves to explore the persecution faced by both. In her *Women Who Run with the Wolves*, psychoanalyst Clarissa Pinkola Estés considers how both wolves and women have been 'hounded, harassed, and falsely imputed to be devouring and devious, overly aggressive, of less value than those who are their detractors'.[83] For Estes, wolves and women alike have been 'predat[ed]' upon 'by those who misunderstand them' in 'strikingly similar' ways,[84] but she urges women to free themselves by embracing the 'wildness' and 'wolfishness' that patriarchal societies constrict and malign.[85]

Wolves have also been reimagined as teachers by numerous authors, such as Elli H. Radinger in *The Wisdom of Wolves* (2019) and Mark Rowlands in *The Philosopher and the Wolf* (2008). These narratives encourage us to discover the meaning of humanity and learn to live better lives by, paradoxically, embracing wolfishness. Although there is a danger of anthropomorphisation and mythmaking of a different kind associated with texts like these (which, by nature, focus on more palatable lupine traits and behaviours, unlike Dahl's Mr Fox), narratives that suggest that we all could learn a thing or two from wolves challenge the perception that they are evil creatures driven by bloodlust and greed. Instead, they showcase wolfish traits that have been and are still often downplayed or ignored: the capacity for love and grief, playfulness, compassion, grace, familial loyalty, patience, teamwork, cooperation, and intelligence. Perhaps most importantly, drawing connections between humans and wolves in positive ways makes them seem less alien, more familiar, and ultimately less frightening than thousands of years of scaremongering and demonisation have led us to believe.

The Wolf Observed

Although stories have been told about wolves for millennia, everything we know about them from a scientific standpoint has been recorded only in the past 80 years.

It was not until the 1940s that wolves were studied in their natural habitat, when American scientist Adolph Murie was sent by the US National Park Service to study the local ecology of Alaska's McKinley Park (now Denali Park). In particular, he was tasked with discovering how wolf predation affected numbers of Dall sheep. Historically, the National Park Service had frequently employed predator eradication as a conservation strategy, not only to protect domestic animals but also to preserve wild ungulate species. This was essentially 'a livestock concept of wildlife administration', with deer considered 'valuable assets' both economically and aesthetically, an important part of the human conception of what 'wilderness' should be.[86] Parks were envisioned as 'landscapes of Romantic leisure, with pastoral scenery, rustic lodges, and vast herds of appealing herbivores for the visiting public to see'.[87] But deer were drastically decreasing in number, fostering even more resentment towards the wolves that hunted them, especially since human hunting was heavily restricted.[88] Wolves threatened these 'assets' and the 'wilderness' that the Park Service was trying to create, and so they were killed.

Wolves were about as well accepted outside the national parks. In the early twentieth century, the Division of Biological Survey (an ancestral organisation of the United States Fish and Wildlife Service) included the destruction of wolves and other predators in its mission, and was allocated the modern equivalent of several million dollars for 'destroying wolves, coyotes and other animals injurious to agriculture and animal husbandry'.[89] While some scientists did object to this approach, one particularly influential biologist employed by the Division of Biological Survey, Edward Goldman, concluded a paper on 'the balance of nature' by arguing that 'large predatory mammals, destructive to livestock and to game, no longer have a place in our advancing civilization', and elsewhere described them as 'a menace to human life'.[90] In 1927, the US federal government themselves proclaimed that they were 'wag[ing] a constant warfare upon predatory animals'.[91] As a result of the government's anti-predator stance, by the 1930s wolves had all but disappeared across the lower 48 states.

But the effect of removing predators was soon to become clear. Populations of deer collapsed as their numbers increased unchecked and they depleted their own food.[92] Ecologists were beginning to understand that wolves and other predators did not hunt their prey to extinction.[93] Predators suddenly became 'special charges' of the National Parks, and the same park rangers who had hunted wolves to local extinction were now schooled in the importance of these animals. All too late, the Park Service celebrated that 'impartial scientific data' had overcome 'ancestral prejudice'.[94]

Yet opinion was slow to change. Murie's study of the wolves and sheep in Mount McKinley Park was prompted by public pressure to kill the predators, which were feared to be reducing numbers of the sheep and other animals. Even a former director of the National Park Service, Horace Albright, wrote in 1939 that 'he found it "very difficult" to accept the idea of protecting McKinley's wolf population in the "territory of the beautiful Dall sheep"', deeming it a '"grave risk" to spend so much time and effort caring for predators' which he felt in fact '"[did] not or need not fall on the National Park Service at all"'.[95] Albright was proved wrong. Murie's study confirmed that the wolves of McKinley Park actually kept the Dall sheep population healthy by preying on its weaker members, and that there was little point in culling predators to increase numbers of their prey. Murie also helped to reframe wolves not as villains but simply as animals, emphasising that packs were family units who treated each other with kindness and friendliness, and not the 'random, slavering mass' they had previously been understood to be.[96] Slowly but surely, in no small part because of the efforts of Murie and other biologists, predator eradication came to an end.[97] They irrevocably changed the narrative among those charged with managing wolves, who had for so long been influenced by the animal's cultural reputation rather than its biological reality.

One such convert was Aldo Leopold, an employee of the US Forestry Service in the 1940s, whose job was to exterminate predators in New Mexico. Leopold documented his transition away from the typical conservationist mindset of predator extermination with a now-famous anecdote about a wolf he once shot:

> We reached the old wolf in time to watch a fierce green fire dying in her eyes. I realized then, and have known ever since, that there was something new to me in those eyes—something known only to her and to the mountain. I was young then, and full of trigger-itch; I thought that because fewer wolves meant more deer, that no wolves would mean hunters' paradise. But after seeing the green fire die, I sensed that neither the wolf nor the mountain agreed with such a view. [...] Since then I have lived to see state after state extirpate its wolves. I have watched the face of many a newly wolfless mountain, and seen the south-facing slopes wrinkle with a maze of new deer trails. I have seen every edible bush and seedling browsed, first to anaemic desuetude, and then to death. [...] I now suspect that just as a deer herd lives in mortal fear of its wolves, so does a mountain live in mortal fear of its deer.[98]

This passage, appearing in Leopold's collected essays on the American landscape and the human-nature relationship, *A Sand County Almanac* (1949), has become a cornerstone of the debate about wolves.

But at the time, Leopold received much criticism by those who, facing continued hunting restrictions, felt that he was simply 'saving the deer in order to feed them to the wolves'.[99] It was not until almost two decades after the publication of *A Sand County Almanac* that the notion that wolves were not malignant tumours sucking the life out of landscapes reached a wider audience, with the publication of Farley Mowat's 1963 book *Never Cry Wolf*. A semi-fictional account of his time studying wolves and caribou in Canada in the late 1940s (Mowat himself noted in 2012 that he 'never let facts get in the way of a good story' and 'was writing subjective non-fiction all along'),[100] *Never Cry Wolf* popularised a more sympathetic approach to wolves both in Canada and worldwide. Mowat described being sent by the Canadian government to the northern wilderness to study wolves and caribou, ostensibly with the aim of proving that wolves, not human hunters, were decimating the deer population.[101] But after observing a wolf family for an extensive period, Mowat soon learnt that 'the centuries-old and universally accepted human concept of wolf character was a palpable lie'.[102] Instead, he found the wolves he observed to be far more intelligent and emotionally complex, and far less bloodthirsty than he had been led to believe by popular (and popular scientific) opinion.

Although it was not entirely credible, *Never Cry Wolf* was an exercise not in rigorous and objective scientific endeavour but in emotional connection, familiarisation, empathy, and in meeting wolves on their own terms. This was a radical departure from the prevailing approach of the Canadian government's biological services division when Mowat undertook his expedition which, as in the United States, was one of extermination.[103] Mowat highlighted that wolves had long been persecuted by a far more bloodthirsty species, *Homo sapiens*, under false pretences of viciousness and danger. Humans had falsely criminalised wolves as indiscriminate killing machines who delighted in blood-sport, as villains who perpetrated mass genocide by annihilating wild deer populations (a charge that Mowat instead levelled against humans), and as vectors of disease.

While *Never Cry Wolf* was particularly impactful in Canada, where the Wildlife Service received a 'deluge of letters [...] from concerned citizens opposing the killing of wolves',[104] Mowat's reconsideration of wolves as intelligent, emotional, playful, loving, and familial, as vital to the landscapes they inhabited, and as the victims of a foolish campaign of destruction, reached all corners of the world. He ultimately helped to rehabilitate the wolf's image on a wider scale than anyone had previously – and, arguably, have since – achieved, for the first time framing wolves as a species worth protecting in their own right.

While there is much to celebrate about the changing narrative surrounding wolves, that narrative remains a story constructed by people. Just as the wolf in the past was (and often still is) mythologised as an evil, bloodthirsty creature, it is now equally mythologised as 'a saintly figure with exemplary morals'.[105] This saintly wolf is not a predator of deer but one that lives almost entirely on mice (one of Mowat's fictions),[106] a benign animal that would never inflict harm on a human (it was often said that there were no documented cases of a healthy wild wolf killing a human in North America in the twentieth century, a claim that may have been true at the time but which was 'quickly interpreted by advocates of wolf conservation as an assertion that wolves simply do not kill people').[107]

Myths about wolves – good, bad, and everything in between – crescendo into a concert of misinformation, so loud that the 'real' wolf is drowned out entirely. Perhaps most ubiquitous is the false notion that wolf packs are organised in a strict hierarchy determined by dominance and submission behaviours (particularly over food), led by the 'alpha female' and the 'alpha male'. This behaviour was first observed in a captive pack of unrelated wolves by researcher Rudolph Schenkel in the 1940s, and the notion of 'alpha' wolves was subsequently popularised. But this dominance hierarchy turned out to be the exception rather than the rule, and was later found to be inapplicable to the family unit that most wild packs comprise.[108] It is about as appropriate to call your own mother and father the 'alpha female and male' as it is to refer to wolf parents in these terms. Yet the myth of the dominance hierarchy and of 'alpha wolves' has been slow to die outside of academic circles, not least because it fits the aggressive, savage picture of wolves that Western culture has painted.[109]

These myths surrounding pack dynamics inform an image that has widely circulated online, which captures a large wolf pack travelling through a snowy Yellowstone landscape in a snaking single file. Three wolves at the head, so the caption says, 'are the old or sick' who 'give the pace to the entire pack', both so that they are not left behind but also so that they can be sacrificed to any attacker who might confront the pack. Five wolves that follow are the 'strong ones, the front line', followed by the 'rest of the pack members', and another '5 strongest following'. A single wolf who trails at the back is supposedly 'the alpha', who 'controls everything from the rear'. None of this is true. But comforting, 'wholesome' myths like this are well suited to the ecosystems of social media.

The ubiquitousness of this image online is testament to the cold truth that mythologising wolves seems to be deeply and irrevocably ingrained in so many of us, whether we are pro- or anti-wolf. Those who hate wolves perpetuate myths about their ravenousness, bloodthirstiness, and their penchant for killing for fun. Wolf advocates, on the other hand, may aim to dispel such notions but in the process will often disseminate misinformation without questioning its veracity, or else 'eagerly seize on any [scientific] study they consider favorable to wolves'.[110]

Yet such studies in turn can be biased in favour of wolves, particularly when it comes to the ecological effects of their restoration. After wolves were reintroduced to Yellowstone National Park in 1995, where the absence of apex predators had suppressed tree growth and regeneration,[111] some biologists were quick to conclude that the wolves had initiated a trophic cascade.[112] Wolves were credited with indirectly affecting numerous species of flora and fauna in the park, from coyotes and deer right down to beetles, and were even said to have improved water quality and reduced its temperature. The media soon publicised – and, in some cases, exaggerated – the results of these studies, 'eager [...] to hype [the] findings'[113] but proving far less enthusiastic about covering any subsequent studies which challenged these conclusions. As a result of this bias, as wolf biologist David Mech noted, a new imagined wolf was created which 'may be just as erroneous' as 'the animal's public image a century ago'.[114]

One of the most famous examples of this misguided narrative is the YouTube video 'How Wolves Change Rivers', uploaded by the Sustainable Human channel in 2014 and since garnering tens of millions of views. This video claims that the reintroduction of wolves created a trickle-down effect which saw them change the behaviour of deer, which allowed trees to regenerate, which caused birds to move in and beavers to arrive and start building dams, which in turn provided habitat for otters, muskrats, ducks, fish, reptiles, and amphibians. Wolf predation on coyotes was also said to allow numbers of mice and rabbits to rebound, in turn increasing numbers of birds of prey, weasels, foxes, and badgers. An increase in bear numbers was attributed to their scavenging of wolf kills, as well as increased availability of berries from the regenerated trees. The *pièce de résistance* of the video, however, is the claim that 'the wolves changed the behaviour of the rivers' because the regenerated trees stabilised riverbanks, in turn causing less erosion, narrowing the water channels, and creating more pools.

It would be nice to believe that this was all true. There could certainly be some degree of accuracy to it – a 2005 study of wolves in Canada's Banff National Park has indeed 'provided seemingly irrefutable evidence of a true trophic cascade from wolves through prey, vegetation and song birds'. But, as Mech points out, this is the only study to do so.[115] It is extremely difficult for scientists to confirm beyond all doubt that the return of wolves affects any species other than those they prey upon,[116] and it is all but impossible to say whether most of the changes that happened in the Yellowstone ecosystem were attributable to the reintroduced wolves. Correlation is not causation – ecosystems are vastly complex and utterly unpredictable.

Even the decreases in Yellowstone elk numbers and the changes in their behaviour seen after the return of wolves could be attributable to myriad factors. There were harsh winters and drought. People hunted enthusiastically, killing almost twice as many elk than wolves between 1995 and 2011. Numbers of cougars and bears increased. All of this, as well as the renewed presence of wolves, affected the elk in different ways.[117] Wolves were a part of it all, but they were far from the only actors in this play. To say that 'wolves change rivers' is a simplification of a much greater picture, another myth that has taken on a life of its own.

This is not to say that the loss of wolves and of the process of apex predation does not impact ecosystems in negative ways, particularly due to the increase in deer numbers that naturally results from the absence of predators, which leads to unsustainable grazing pressure and, in turn, has knock-on effects for the rest of the ecosystem. Eradicating predators results in an 'inexorable' march towards 'ecosystem simplification accompanied by a rush of extinctions'.[118]

But putting wolves back is not a magical fix to overhaul every part of a degraded ecosystem, like pushing fallen dominos back the other way. Each ecosystem is a complex and intricate web; threads snap and change its shape, but weaving the loose ends back into the tapestry does not necessarily mean that the picture will look the same as it once did.[119] Reintroducing predators is just one – albeit highly important – part of nature conservation, much of which also depends on changing human uses of the natural world. Environmental restoration simply does not have a single, miraculous answer.

Wolves are one of many crucial components that contribute to the health of the ecosystems to which they are native, but revering them as 'saviours' of ecosystems only invites scrutiny, backlash, and persecution if and when they do not magically restore a landscape.[120] As David Mech puts it, the wolf 'remains as one more species in a vast complex of creatures interacting the way they always have', playing an integral role in the web of life.[121] But as with the negative images of the wolf that have proliferated throughout history, the canonising of the wolf as a saintly figure whose effect on ecosystems is nothing short of miraculous equally fails to do the animal justice. In short, in Mech's infamous words: the wolf 'is neither saint nor sinner except to those who want to make it so'.[122]

CHAPTER 7

Rewolfing

The wolf has had a long journey. Buffeted around as her crate was lifted from vehicle to vehicle, turbulent in a cold hold above the North Sea, deposited on tarmac amid a cacophony of engines, into the back of yet another truck, swaying and jolting.

Then, the movement stops.

She is frightened.

The inside of her grey box is dimpled with dim light which trickles in through small holes in its sides. There are many sounds. Not only the mournful wind and the fizz of light rain pattering on the roof, but a hum of voices, moving feet, a curious mechanical clicking, rustling coats keeping cold skin dry. The scents are metallic and wooden within, but beyond her four walls she smells the rain, the petrichor it has created, coldness, grass, alder, pine, rodents, deer, birds, water, chemicals, plastic, humans. Other wolves, both known and unknown, each within its own box.

All at once, one side of her box shoots up. She starts backwards at the sudden movement, eyes wide, ears pressed to her skull, before peering cautiously out. She sniffs, tentatively placing one paw on the sodden grass.

The next moment she bursts forth, a streaking flash of fur dashing towards the trees. Powerful legs propel her swiftly away from cameras and spectators. Her fur glints in a sudden burst of sunshine revealed by a break in the clouds on this dreich November day. Undertones of red gleam briefly before she darts into the trees and is enveloped by shadow and branch.

She does not know it, but she is the first of her kind to set foot here in over 300 years.

Wolf Land has come alive once more.

Oh Deer

One of the biggest problems faced by many wildlife managers across Britain and Ireland walks on hooved legs. It spends its days in woodland and on the moors, and it comes in six species of varying sizes. It makes conserving and restoring woodland a logistical and financial nightmare at best, or an utter failure at worst. This problem is called deer.

Like wolves, deer are an important part of the ecosystems to which they are native. They create and maintain open areas through browsing, grazing, and disturbing the earth, providing homes for the plants and animals (many endangered or threatened) that prefer these types of habitats. They recycle nutrients which improve the soil, disperse seeds with their dung, and provide essential food for scavengers when they die.

But without any predators, save humans, deer numbers have increased to at least several million – their highest for perhaps a thousand years.[1] Record numbers are culled each year, but herds still number in the hundreds or even thousands, marching further and further across the map as they expand their empires. These 'unchecked megaherds' are anything but natural,[2] nor is the presence of three invasive, non-native species – sika, muntjac, and Chinese water deer.[3]

The explosion in deer numbers has caused a multitude of problems. They eat and trample farmers' crops, as well as saplings in plantations and commercial woodlands, causing millions of pounds-worth of damage and making a mockery of the millions that are spent on their management each year. They also compete for grazing with sheep, and spread diseases and parasites which can infect livestock. Even more costly for both humans and deer are the tens of thousands of traffic collisions that they are involved in across Britain and Ireland each year, which are expensive at best and dangerous or deadly at worst. These incidents are becoming more and more frequent as both deer and car numbers increase and as deer ranges expand to encompass more urbanised areas. They also contribute significantly to the spread of Lyme disease, which infects thousands of people across Britain and Ireland every year, who are bitten by members of the thriving tick population that the deer host. They have even been known to attack pets and people.[4]

Forests are hard pressed to regenerate when young trees are nibbled by hungry cervid mouths, flattened by their hooves, and stripped of bark by their antlers and teeth. If saplings are lucky enough to survive – and many are not – their growth is often stunted. Lower-growing plants are likewise trimmed and trampled so severely by these four-legged lawnmowers that they cannot flower and set seed, depleting numbers of important plants such as bluebell, and rarities such as oxlip and certain orchids. Instead, they leave behind an understorey dominated by grass and bracken, which supports far fewer species.[5] The knock-on effects are felt throughout every trophic level of the much-depleted woodland ecosystems,[6] with overgrazing reducing important sources of food and shelter for many mammals, birds, and insects.[7]

Trying to plant trees in areas of high deer density is like trying to build a house on sand. Upland woodland is thought to be able to regenerate when deer live at densities of no more than 7 per square kilometre. But in Scotland, deer densities average around 11 to 12 per km^2 and in some places numbers may reach as high as 25 to 50 per km^2, or even 150 per km^2 during winter.[8] Though there are ways to help establish saplings, like tubing and fencing, these are costly (tubing can multiply the cost of planting a single tree eightfold), frequently ineffective, and come with their own problems – fences fragment habitats and tubing often leaves the trees it protects less stable.[9]

Hundreds of thousands of deer are slaughtered across Britain and Ireland each year in an attempt to keep their numbers in check, but even so much bloodshed is ineffective against the ever-advancing tide of hooves. It is a financial sinkhole for landowners and managers who, to add insult to injury, make little money from the venison they produce. It is estimated that half of the deer population in Britain would have to be culled every year to bring numbers back down to natural levels.[10]

Bringing Back the Wolf

In the context of the British and Irish deer problem, many are beginning to seriously consider whether in fact wolves could and should be brought back.

Predation by large carnivores is a unique ecological process which is integral to biodiversity and ecosystem function. But this process is entirely absent in Britain and Ireland. Humans cannot replace this lost process no matter how many deer we kill, because it is not merely an issue of numbers but of prey behaviour and the transfer of nutrients throughout the ecosystem.[11] The landscape evolved alongside wolves and other large carnivores, and it is now feeling their absence.

There is no doubt that wolves would affect the deer populations upon which they predated in some way (red deer would likely be the primary prey of a reintroduced wolf population, though they would also probably take roe deer, wild boar, and other smaller animals).[12] Quite simply, with an average-sized wolf needing a minimum of around 1.8kg of meat per day,[13] they would have to kill deer to eat. But how exactly this would affect deer numbers is another question.

Wolves can only control deer numbers if there are enough of them making sufficient kills to outpace the deer population's rate of reproduction. But these predators typically live at fairly low densities (though smaller areas can support more wolves if they also house abundant prey), and because deer are large, wolves need to kill fewer of them to sustain themselves than they do smaller animals.[14] A 1992 study of an area of Spain in which red deer lived at densities of 10–20 individuals per square kilometre suggested that even if the local wolf population ate nothing but these deer, they would only remove about 1% of the species' total biomass.[15]

But wolves do not just affect deer on a numerical basis. Some prey species are capable of sniffing out carnivores and even identifying them by their spoor and urine, and they use this information to alter their behaviour. Caribou, for instance, use these signs to ascertain the numbers of wolves in an area, and they adjust their calving behaviour to reduce the risk of losing their young to predation.[16] Research has also suggested that fear of predation can affect deer nutrition and body condition,[17] as well as reproduction success, offspring survival, and therefore population growth.[18] It can even change the neurobiology of the individuals who escape life-threatening encounters.[19]

It is thought that the behaviour of deer is also affected more generally by the presence of predators in an ecosystem, who create what is known as a 'landscape of fear'. This theory suggests that when predators are present, deer avoid or spend less time in areas where they are most at risk (such as places where it would be more difficult to escape from), allowing vegetation to rebound in these areas. Even in 'safer' areas, it is suggested that deer spend less time foraging because they have to remain vigilant, reducing grazing pressure in these places as well. One study of Poland's Białowieża Primeval Forest indicated that deer browsing was less pronounced in 'core areas' of wolf territories, especially in places from which it would be difficult to escape, allowing almost 60% more saplings to grow above deer-browsing height than outside of these areas.[20] Numerous studies have also documented increases in tree and plant height in parts of Yellowstone National Park since the reintroduction of wolves, which has famously been interpreted as evidence of trophic cascades.[21]

But this effect has not been seen everywhere. One study found that while Yellowstone elk do respond to wolf presence by being more vigilant and by spending more time on the move, they do not drastically change their eating habits or their use of the habitat.[22] And although trees and plants have increased in height in some parts of Yellowstone, in others – even 'high-risk areas', which elk can simply access when wolves are resting – they have not, with many plants still being heavily grazed despite large decreases in elk numbers.[23] One study found that wolves and elk only actually encountered each other on average once every nine days.[24] There's also safety in numbers – when deer herds are larger, each individual can afford to be less vigilant and can spend more time foraging.[25]

In Europe, meanwhile, where wolves have been slowly but surely recolonising their former ranges, there is little concrete evidence of widespread ecological change due to wolf predation of overpopulated deer. While research from France and Poland has drawn correlations between wolf presence and ungulate numbers, as well as grazing intensity,[26] one study of wolves and moose in Scandinavia found that moose numbers actually increased in wolf territories, leading to increased grazing pressure on Scots pine in these areas. This unexpected effect was perhaps attributable to the fact that wolves selected habitats based on prey abundance – they settled in areas with moose populations so large that the wolves could not offset their growing numbers. It may also be that moose numbers increased because human hunters altered their behaviour, shooting fewer moose in anticipation of population declines caused by wolves.[27]

We simply cannot know how the reintroduction of wolves to Britain or Ireland would affect the deer, not only in terms of numbers but especially in how deer would adapt their use of and distribution within the landscape, and how it would impact upon their reproductive success. Unlike in Yellowstone, where elk coexisted with other predators (albeit in reduced numbers) when wolves were absent, deer in Britain and Ireland have not lived alongside an apex carnivore for centuries. It is thought that the defensive behaviour of prey species changes when predators become absent, and British and Irish deer have had several hundred years to adapt to life without fear of wolves or other predators. How would they react when wolves came back? Would old instincts be reawakened, or would they need to relearn their fear of their four-legged foe?[28]

Although research has been conducted into the potential outcomes of wolf reintroduction to Scotland, predictions from each study vary widely. One 2007 simulation found that a population of wolves reintroduced to the Scottish Highlands beginning with just 12 individuals would reach an equilibrium of 25 wolves per 1,000 square kilometres. These animals, the model suggested, would lower the density of deer from more than 20 per square kilometre to an ecologically sustainable level of 7 per square kilometre within 60 years, which could lead to forest regeneration.[29] A 2025 study modelled the outcome of wolf reintroduction to four of the largest Wild Land Areas in the Scottish Highlands, an area totalling 12,167 square kilometres which, the study suggested, could house a population of around 167 wolves living at densities of 13–14 individuals per 1,000km^2. Within 20 or so years of wolf reintroduction, it was estimated that deer densities in these areas would decline from 9.35 per km^2 to 4 per km^2, allowing woodland to recover and expand. The researchers calculated that these expanded woodlands would sequester around one million tons of carbon dioxide per year, worth more than £25 million annually based on carbon pricing.[30]

But in contrast, a 2019 study found that even if a wolf population in the Scottish Highlands was large enough to be self-sustaining, it is unlikely that densities would ever to be high enough to regulate deer numbers. This research suggested that wolf densities of more than 80 individuals per 1,000km^2 would be required to exert any influence on deer numbers.[31] With wolves very rarely living at densities of more than 50 individuals per 1,000 square kilometres,[32] this number is too high to be realistic. Another study from 2007 similarly found that an area of 1,000km^2 in the east of Scotland housing around 23 deer per square kilometre could support a population of 43 wolves, but that these wolves would not kill anywhere near enough deer to suppress the population.[33] A 2009 study, on the other hand, suggested that wolves could potentially create a localised 'landscape of fear' among deer in Scotland, even if they did not alter prey densities. The researchers posited that this would allow forest to regenerate in places avoided by deer,[34] creating a patchwork of both open and wooded habitats across the landscape.[35]

But although wolves would not necessarily 'fix' the deer problem, they would facilitate other natural processes that are essential to healthy and diverse ecosystems. Wolf kills do not only feed wolves – they provide sustenance for a variety of mammal, bird, and insect species, as well as bacteria, with nutrients ultimately recycled into the soil. In Yellowstone, more than ten species have been observed taking advantage of wolf kills, providing 'food for the masses' which may be essential for the survival and reproduction of such species.[36] As with natural predation, humans cannot replicate this ecological role because deer removed by culling and hunting are transported from where they fall, ultimately leading to a loss of nutrients from the ecosystem.[37]

Wolves may also limit numbers of smaller mammalian and avian predators such as foxes and crows, species known as 'mesopredators'. It has been theorised that the extirpation of an ecosystem's large carnivores causes a phenomenon known as 'mesopredator release', whereby the absence of apex predators allows smaller carnivores to expand in number and range, with knock-on effects for their wide range of prey species, including birds, small mammals, reptiles, and amphibians. Though this theory is unproven, the UK does have larger populations of mesopredators than many other countries in Europe.[38] Returning apex predators could theoretically limit numbers of foxes and badgers through competition and through direct killing. As with the deer overpopulation, however, it is not a foregone conclusion that the return of wolves would counterbalance increased mesopredator numbers,[39] especially since human activities have also impacted the numbers and range of both mesopredators and their prey.

But just because they are not a magic bullet to solve all of our conservation problems, this does not make wolves any less important to the ecosystem – an ecosystem that evolved along with them, and they with it.[40] Though deer would still need intensive management for the foreseeable future (and beyond, in places unsuitable for returning wolves), restoring the 'crucial and irreplaceable' natural process of apex predation would be a significant step towards creating healthier ecosystems.[41] On the other hand, if we 'fail to restore keystone interactions' such as predation, we 'risk a wave of secondary extinctions'.[42]

Wolves can also benefit people in more material ways. Their charisma and beauty would undoubtedly make wolves a huge tourist draw, not only for opportunities to see the animals in the wild (for which the odds are always slim), but because even to walk in wolf country, in a place bold enough to be the first in Europe to reintroduce wolves, would be a thrill for many. Thousands of people visit Germany's Harz National Park, where lynx were reintroduced at the start of the millennium, not on the promise of seeing these secretive animals but simply because knowing that they are there, prowling silently through the forest, is a draw in itself.

There is already great precedent for wildlife tourism in the UK. Whale-watching in western Scotland attracts almost a quarter of a million people each year, while the Scottish osprey population draws almost 300,000 visitors to watch points annually.[43] Hundreds of thousands of people even travel to Scotland for the legendary animal that lives beneath the waters of Loch Ness, an iconic monster worth tens of millions of pounds to the Scottish economy. Wolves could become an icon of the Highlands, as synonymous with Scotland as Nessie or the Monarch of the Glen.

Wolves in Yellowstone bring tens of millions of dollars to the Greater Yellowstone area every year. One study estimated that the economic benefits of wolves in nearby Washington state outweighed the management costs by more than 150 times.[44] Closer to home, reintroduced white-tailed eagles bring several million pounds to the Isle of Mull each year, with tourism surrounding the raptors creating more than a hundred job opportunities. Farmers in this area, who could have been negatively affected by eagle restoration due to predation on lambs, have instead capitalised on increased tourist numbers by letting out accommodation and hosting wildlife tours on their land, adapting and thriving in unexpected ways.[45]

There is also a moral argument. As is stated in the manifesto of the International Union for the Conservation of Nature (IUCN) Wolf Working Group (previously known as the Wolf Specialist Group), 'wolves, like all other wildlife, have a right to exist in a wild state'. They have an intrinsic value which 'is in no way related to their known value to mankind', but instead 'derives from the right of all living creatures to co-exist with man as part of the natural ecosystems'.[46]

Wolves have a right to live in the places that they called home for hundreds of thousands of years, no more or less than humans, deer, foxes, pine martens, beavers, or any of the thousands of species that live in Britain and Ireland. And as the species that removed them, perhaps it is our duty and our moral obligation to return them – to right the wrongs which humans inflicted upon both the ecosystem as a whole and upon the species that we relentlessly persecuted in pursuit of our own prosperity and because of our own culturally ingrained hatred and fear.

We have a duty to put our money where our mouth is. We should make the same 'noble, self-sacrificing conservation efforts' that we request of communities in developing countries who live alongside animals like lions and tigers, which kill hundreds of people every year, are far more challenging to share the landscape with, and which significantly affect people's livelihoods.[47] All while we refuse to live alongside far less dangerous predators because we deem it too difficult.

Welcoming the Wolf Home

Accepting wolves is far easier in theory than when they are actually at the door.[48] The question is not simply 'could wolves live here?', but 'could we live with them?'. Are we ready to walk among wolves once more?

The IUCN requires 'strong evidence that the threat(s) that caused any previous extinction have been correctly identified and removed or sufficiently reduced' before a reintroduction occurs.[49] In the case of the wolf, this is a thorny issue. Wolves died out because of persecution resulting from both real-world conflict and the cultural baggage attached to them, both of which could still affect the success of a reintroduction project. The idea of bringing wolves back is, for many, an uncomfortable one. Logic and reason can inform us that wolves are less dangerous than cows, dogs, and driving, but emotions are powerful clouds of judgement. If wolves are to return, we must find ways to mitigate these conflicts and to lighten the cultural baggage, not only in the lead-up to their return but for many years afterwards.[50]

We have entirely forgotten how to live with wolves. Not only is it not within living memory, but it was not even within the living memory of our grandparents' grandparents. We barely know how to safely walk through a field of cows, let alone a forest that is home to lynx or wolves. While people in other countries around the world coexist with not only wolves but bears, lynx, lions, tigers, hippos, hyaenas, sharks, crocodiles, and poisonous snakes and insects, we struggle to share our landscape with foxes, badgers, beavers, and birds of prey, often in an uneasy détente at best. Animals and plants are frequently vilified in the press where, as conservationist Hugh Webster puts it, 'innocuous encounters' are 'dramatise[d] [...] to stir up headline-grabbing hysteria', exacerbating a fear of wildlife that is a product of (and reinforces) our disconnection from the natural world.[51] Seagulls 'take over' entire streets and are 'a danger to society', leaving children 'living in fear'. Flying ants 'invade' Britain like a winged scourge, and 'hungry kites' 'attack' towns.[52] When wild boar returned to England the press referred to them as 'killer[s]' and 'vicious' 'monsters', raising concerns that these animals would be a danger to humans and livestock, despite unprovoked attacks on humans and predation on livestock being entirely unheard of.[53]

It is little wonder that the UK and Ireland have extirpated (and failed to restore) more animal species than any other European countries.[54] Though we have managed to reinstate some previously extinct species, perhaps most notably the beaver and the white-tailed eagle, the reintroductions of both were – and remain – complicated and controversial. Conflict has led to hundreds of beavers being killed in Scotland since their return, with over 200 licences to kill handed out by the Scottish government even after beavers became a protected species. More have certainly been killed illegally. Conflict with gamekeepers and farmers has likewise led to numerous white-tailed eagles being shot and poisoned in Ireland, Scotland, and England since their return.

In a place where, until recently, foxes have been terrified, chased down, and torn apart by dogs, is there any hope for the wolf, the fox's larger and far more controversial counterpart? Would it be right to introduce wolves into this quagmire of fear, disconnect, and intolerance? Even the reintroduction of lynx, a secretive, shy, and relatively small animal with little cultural reputation to speak of, is contentious. Wolves, on the other hand, are up against thousands of years of fictionalising.

Stories both reflect and affect attitudes towards wolves. Opinions, beliefs, ideas, myths, and all the trappings of cultural baggage – whether good or bad, realistic or wildly erroneous – have a profound impact upon whether wolves thrive, survive, or die. Whether or not we have actually encountered these animals in real life, or whether we even know much about them, most of us have an opinion. It is impossible not to. Stories kept our fears alive and howling long after we purged wolves from the landscape. We live in the looming shadow of the Big Bad Wolf who huffs, puffs, and blows the house down – a fitting symbol of the threat wolves are often thought to pose to civilisation itself. Our stories live among us far more vividly than the wolf that lives in the real world, who is safely relegated to the distant 'elsewhere' of other nations. Biological reality has been supplanted by cultural fantasy. We do not fear wolves: we fear what we have made them.

But these animals are far less deadly than our fictions have made them seem. Tales of man-eating wolves prowling around flimsy wooden carriages trapped in deep snow are as old as time. But fairy tale wolves and all those things that go bump in the night, whose eyes flicker in the dim light and who snarl at the edge of the fire – they exist to make stories interesting. The source text, the biological animal, is buried beneath such creations to the point of suffocation.

Even in the distant past, when there were far more wolves and far more people living in rural settings alongside them, attacks were uncommon.[55] Today, the factors that led to such attacks have long since declined in the Western world.[56] Fewer people live in rural areas, and livestock are no longer entrusted to children. There is also abundant wild prey (in times and places where wild prey were depleted, wolves naturally turned to domestic species to sustain themselves, bringing them into closer contact with people), and persecution may have led to natural selection in favour wolves who are more shy and wary of humans.[57] Rabies, the cause of the vast majority of wolf attacks, is no longer prevalent across much of Europe.[58] On the Continent, where there were approximately 21,500 wolves as of 2022, you are more likely to be struck by lightning than attacked by a wolf.[59] Likewise, despite housing a population of more than 50,000 wolves, attacks across the entire North American continent between 2002 and 2020 numbered in the single digits, only two of which have been fatal.[60]

Researchers enter wolf dens to remove pups even while the parents are present.[61] They spend countless hours in extremely close proximity to packs – as near as one metre – and handle wolves that are not sedated, all without issue.[62] David Mech recalls no less than a dozen summers spent 'virtually living with a pack of wolves in the high Arctic', with 'only the thin nylon' of his tent walls to serve as a barrier between him and the animals. The biggest problem they caused was rolling about on his newly cleaned undergarments.[63] Most wolves do not view or treat humans as prey,[64] and predatory attacks on people are 'much more a part of the "normal" behaviour of other large carnivores (bears, cougars, tigers) than that of wolves'.[65]

On those rare occasions when wolves have attacked people it is often humans who cast the first stone, forcing wolves to act in defence of themselves or their pups. Most such instances involve a single bite as a means to distract the assailant so that the wolf can escape.[66] For example, an incident in Spain saw a shepherd digging out a wolf den to remove the pups. His dogs cornered the mother wolf while the shepherd threw rocks at her, and she bit him before running away.[67] It is a minority of wolves, even in such circumstances, that will attack.[68]

People live in and venture into wolf country all around the world, and many of us in Britain and Ireland visit destinations that have significant wolf populations without even realising it.[69] There is a reason why seeing a wolf in the wild is a once-in-a-lifetime experience. Although wolves are capable of living in close proximity to towns and cities, they are usually very wary of humans,[70] with even the most experienced wolf researchers frequently struggling to catch a glimpse of their subject without extensive tracking, camera traps, GPS collars, or aircrafts. In areas that they share with humans, wolves will generally stay away from people as much as possible, avoiding settlements and seeking out places where humans venture less often.[71] When approached by people, wolves will almost always retreat before the two-legged interloper is even close enough to see them, and will be long gone within an hour or so.[72]

Wolves can sometimes become less wary of people, a phenomenon known as 'habituation'. Habituation can occasionally lead to attacks, often when wolves have come to associate humans with food (such as when people have fed them, teaching them that being bold around humans is rewarding).[73] Generally, however, habituation is associated with tolerance of human presence rather than aggression or an entirely lost fear of people.[74] In Yellowstone National Park, which is visited by several million people each year, wolves generally tolerate but are still cautious of humans. Since their reintroduction, only 55 wolves have approached people or proved bold enough to even stand their ground if people approached them. Of these individuals, 17 never showed such boldness again, 38 were 'hazed' (an 'aversive conditioning' technique by which wolves are taught to fear people, using methods such as subjecting them to loud noises and shooting them with rubber bullets), and two were killed as a precautionary measure because they had begun to associate people with food. None of these wolves attacked a person.[75]

Wolves are certainly not the man-eaters that they are so often portrayed as, but nor are they entirely harmless. Living alongside wolves, as with many other wild animals, does not come without risks – as David Mech notes, 'we should view wolves with the same healthy respect due any potentially dangerous animal'.[76] But humans have killed millions more wolves than wolves have killed humans over the centuries, a decidedly 'one-sided war'.[77] As carnivore researcher John Linnell puts it, 'the risks of being attacked by a wolf are not zero, but are clearly so low that they are virtually impossible to quantify, especially when compared to the other background risks associated with living'.[78] In fact, if the statistics of wolf attacks are compared with the data related to other large carnivores, it is clear that despite easily being able to kill a person if they wanted to,[79] 'wolves are among the least dangerous species for their size and predatory potential'.[80]

There would certainly be problems if we reintroduced wolves. But they would not be problems of rabid wolves hunting down children as they walked to school or stalking hapless travellers in the woods. The biggest problem would be wolves being killed for the cultural baggage under which they are laden, for their undeserved reputations as vicious predators of humans.

Learning to Coexist

Predation of sheep, on the other hand, is a legitimate issue.

Although wolves tend to favour wild prey even though domestic species are generally much easier to catch,[81] much of the local sheep population in Britain and Ireland roams freely over hill and dale for most of the year, which is clearly not compatible with wolf reintroduction. Norway is often cited as an exemplar of the dangers of returning wolves in the current climate of sheep farming. In Norway, as in Britain and Ireland, sheep are left to free roam, unprotected and often infrequently checked on during the milder months of the year. The numbers of dead sheep that the Norwegian government pay compensation for as wolf kills are indeed stark. Between 2010 and 2015, a Norwegian population of 33 wolves was held responsible for a staggering 2,037 sheep deaths. But although the conditions are highly conducive to wolf depredation – free-roaming, unsupervised sheep being an easy target – the majority of these livestock deaths were unverified as wolf kills. Just 5–10% of missing sheep are even found and inspected every summer. The remaining 90–95% are assumed to be caused by wolves, lynx, wolverines, or bears, despite the fact that numerous other fates could have befallen the missing sheep, including death due to severe weather, lack of food, injury, and illness.[82] To say that 33 wolves killed over 2,000 sheep in a space of five years is highly misleading.

But this is a highly emotive issue. While factors such as disease, accidents, or freak weather are responsible for the deaths of far more livestock than predation, they are much less controllable. Much less visceral, much less likely to haunt one's nightmares or occupy one's worries. Foxes, for instance, are extremely controversial in Britain and Ireland, being perceived as highly threatening to farmers' livelihoods despite less than 1% of lamb deaths being attributable to them, compared with the 15% of lambs that die of natural causes during or shortly after birth.[83] The more frequent damage to crops caused by small mammals such as rodents, meanwhile, can often be more impactful than the damage caused by predators, and yet it is frequently accepted, tolerated, and considered merely a 'fact of life'.[84] Attacks by predators, on the other hand, often feel intentional and personal.[85] Like foxes, wolves are often seen as highly problematic for livestock production across Europe, despite being responsible for less than 1% of sheep deaths.[86]

Sheep farming is an integral part of Britain and Ireland's natural and cultural heritage. The arrival of livestock in Britain and Ireland in the Neolithic transformed the wooded landscape into a mosaic of closed and open habitats, which may in fact 'have more closely reflected' the environment of the Ice Age interglacials before humans arrived than the widespread forests of the Mesolithic.[87] Being the 'evolutionary context under which Britain's biodiversity evolved and thrived',[88] mixed mosaic habitats are essential to biodiversity today,[89] and agriculture has created and continued to create habitats for numerous species. The 'distinctive flora of chalk downland, for example, was shaped by centuries of continuous grazing by sheep', and human patterns of land management such as coppicing and heather extraction have 'fostered a complex and dynamic landscape, proliferating niches and habitats and opportunities'.[90] Traditionally managed landscapes where sustainable levels of livestock and crops are farmed amid trees and grassland, such as the Iberian *dehesa* or *montado*, are among the most biodiverse in Europe.[91]

For people, sheep were woolly wealth-makers, making their way into our language, our landscapes, our stories, our skills and traditions, our clothes, our stomachs, our history,[92] and now, our very conceptions of what the natural world is and should be. The image of sheep-inhabited, stone-walled patchworks of fields stretching into the distant rolling hills that we inherited from the Romantics is, for many, synonymous with and an integral, quintessential part of life in Britain and Ireland.

But modern sheep farming is largely incongruous with this picture of nostalgic rural idyll. The sheep-dotted landscape may feel timeless, a living heritage harking back hundreds of years, but it is often in fact the inheritance of changes in the agricultural industry which took place less than a hundred years ago.

Large-scale sheep-rearing has characterised Britain and Ireland livestock farming since the Second World War, when an emphasis on self-sufficiency led to the rapid expansion of national flock numbers.[93] Some 20–30 million woolly grass-munchers now call the UK home, one of the biggest national flocks in the world, while the Republic of Ireland houses an impressive 3.5–5 million sheep across its 70,273 square kilometres.[94] Sheep are most prolific in Wales, where they occupy a staggering 64% of the land area, and outnumber people by more than two to one.[95]

Cultures of shepherding, transhumance, traditional hill farming, and mixed farming methods declined in favour of mechanisation, farming on industrial scales, and intensification to maximise yields, which has resulted in 'the homogenization of the landscape' from a mixed mosaic of habitats into vast swathes of open areas.[96] While grazing is an important ecological process which sheep can fulfil at low densities, there are simply too many sheep to be beneficial. Throughout Britain and Ireland sheep overgraze and trample plants, damaging the upland woodland and heath in which they often flock and compacting soil which prevents it from absorbing rain, increasing the likelihood of flooding.[97] They have been associated with declines in brown hare, redshank, and lapwing,[98] and have been deemed even more ecologically detrimental than red deer in some places.[99] Overgrazing has created what *Rebirding* author Benedict MacDonald terms 'a landscape unintelligible to most native wildlife',[100] a landscape that has changed too rapidly for species to adapt. Even putting the ecological impacts to one side, modern sheep-farming has become disproportionately influential compared to the significance of lamb in our diets.[101] And with low employment rates, high land usage, and heavy subsidisation from the taxpayer, the current way in which sheep are farmed is simply unsustainable.

Returning to traditional and non-intensive ways of farming could not only reduce the ecological impact of sheep and help to keep important cultural practices alive,[102] but would also facilitate coexistence with predators. Our ancestors lived and farmed alongside wolves for centuries, and they developed effective ways to minimise livestock losses. These traditional methods have been revived, tried, and tested across the world, and are being implemented alongside modern techniques which capitalise on scientific and technological advancements.

Wolves did not stop people from farming over past millennia, nor do they stop people all over the world today. Reintroducing these predators could be an exciting opportunity to relearn old techniques, reconnect with traditions buried under the sands of time, and tread in the footsteps of our ancestors.[103] As naturalist Lee Schofield notes, 'close shepherding is an important missing component from modern hill farming'. This lost tradition saw shepherds intricately attuned to the landscape, moving sheep around to ensure that no single area was overgrazed, in turn improving the fodder and soil quality and allowing plants to regenerate. These benefits were driven in no small part by the presence of wolves, which necessitated the practice of close shepherding. Thus, Schofield argues, wolves are not only important components of healthy ecosystems, but also 'critical to the maintenance of culturally valued farming practices'.[104]

Some of the easiest-to-implement, lowest-tech methods can significantly reduce depredation simply by making livestock much more difficult to get at. Fences (whether electrified or not, fixed or portable) can keep wolves away, as can fladry, a strange but cheap, low-tech, and portable solution utilising strings of flags which wolves for some reason do not like to cross, and which can be made even more efficient with electrification. Avoiding grazing animals in 'hotspots' such as forests and areas of tall grass can also greatly reduce predation risk, as can keeping livestock within enclosed pens overnight.[105] Farmers in Sweden and Finland lose between 100- and 1,000-times fewer sheep than their Norwegian neighbours simply by keeping sheep away from wolf habitats and instead housing them in enclosed fields.[106]

Even if sheep are free-ranging, they can still be protected using shepherds and livestock guardian animals.[107] Guardian dogs have been employed to protect livestock from wild animals since at least the fourth century BCE;[108] their presence alone can help to deter wolves and prevent them from gaining access.[109] Livestock guardian dogs can also be specifically bred and trained to detect predators and alert farmers to their presence, to actively protect livestock (whether through physical alteration or by engaging in dominance behaviours to keep wolves at bay), and to chase wolves away.[110] Even llamas, donkeys, and cattle can be used to deter wolves when they are integrated into sheep herds, since these animals are protective, territorial, and wary of canids.[111] Farmers can also capitalise on the wolf's fear of humans by patrolling pastures, a practice known in the United States as range riding, which has proved highly effective.[112]

Wolves can also be scared away using light and sound emitters, drones, or by shooting them with rubber bullets, paintballs, or beanbag shells. Electronic collars are even being tested to modify wolf behaviour by negatively associating livestock with an electric shock. One 1970s wolf reintroduction project in Soviet Georgia utilised shock collars to great success, conditioning the wolves prior to their release by creating an association between the discomfort of the shock and the presence of both humans and livestock. Within a month the wolves could not be approached, and there was no wolf–human conflict for almost 20 years after they were released.[113] Aversion can also be a powerful tool for changing behaviour in the long term, particularly by teaching wolves that livestock are not a rewarding food source. Dead livestock can be laced with chemicals to make the wolf sick, while a similar approach involves 'livestock protection collars' which release a chemical to induce vomiting if a predator goes for the wearer's throat.[114] Wolves who experience these depredation prevention measures soon learn to adapt by switching to wild prey.[115] Within a few generations this learned behaviour becomes ingrained in family lines, leading new generations of wolves to ignore livestock entirely.[116]

Combining techniques such as these has proved to be an effective solution to livestock depredation in Idaho. Operating since 2008, the Wood River Wolf Project has found its non-lethal methods of livestock management to be over three times more effective than simply killing wolves. Slashing depredation to just four sheep among tens of thousands, the total number lost to wolves in the operation area is 90% lower than in the rest of the state.[117]

Farmers can also be rewarded for coexisting with predators. The global Wildlife Friendly Enterprise Network offers 'Predator Friendly' certifications for farmers in Italy and North America who act as 'wildlife stewards' by implementing measures to coexist with wolves.[118] These certifications allow farmers to charge more for their produce, generating revenue that can be used to fund livestock protection measures and coexistence programmes.[119] Reindeer herders who allow predators to use their land are also rewarded financially in Sweden, receiving payments when lynx, wolverine, and brown bears reproduce in areas where they graze their herds.[120] Similarly, the Defenders of Wildlife charity rewards ranchers in the USA who have breeding pairs of wolves on their land. Similar schemes have already been used in Scotland, where farmers have been given financial incentives to maintain populations of various goose species on their land.[121]

Farmers should not be expected to pay to implement depredation prevention measures, nor should they be expected to do so without support. Comprehensive compensation schemes for any animals lost to predation must also be put in place as an assurance in case non-lethal methods to prevent depredation prove unsuccessful. As a last resort, wolves who repeatedly attack livestock can be humanely killed, although this is not a quick fix. Research has found that while killing wolves that have depredated on livestock may reduce attacks on the farm in question, other farms nearby may experience increased depredations as a result of pack breakdown, since lone wolves and packs with fewer members will target prey that is easier to hunt alone or in smaller numbers.[122]

Wolves are perhaps one the most divisive animals on the planet. Not only are they hated and loved in the extreme, but they represent larger sociocultural issues of identity, ideology, power, control, culture and tradition, human rights, economics, environmental values, and party politics, reinforcing the lines that we instinctively draw to separate 'us' from 'them'. The debate about reintroducing and conserving wolves is thus deeply political, value-laden, and emotional. As is so often the case when it comes to wolves, it is about so much more than just the animals themselves.

For many people who live and work alongside wolves across Europe, the wolf is a symbol of urban decision-makers forcing their values on the rural population – values with sharp teeth and bloodied claws. A conflict of this nature is already taking place in Scotland, where white-tailed eagles on the Isle of Mull are seen by some as 'interlopers' that have been 'dumped' on farmers by conservationists without sufficient consultation, whereas the golden eagles that never disappeared are 'seen as part of the heritage of the land', despite both species preying on lambs.[123]

For many people living in rural Britain and Ireland, especially farmers, rewilding is a romantic urbanite fantasy in which places perceived as 'empty wildernesses' – but which are in fact cultural landscapes lived and worked in by local people – become the ecological and ideological playgrounds of distant city-dwellers with little understanding or experience of such landscapes, their socio-natural histories, and the role that people have played in shaping them. As author and farmer Patrick Laurie notes, 'ecologists [...] come from somewhere else and they usually tell us we're wrong' or 'can't be trusted with nature', with the labelling of farmers by some rewilding proponents as 'architects of a "sheep-wrecked" wet desert' being particularly alienating.[124] For many rural residents, wolves would feel like an unnatural, unwelcome, and unnecessary burden, a threat to rural life and traditional land uses unjustly imposed by idealistic and detached urban bureaucrats.[125] Jim Crumley, who advocates for the wolf's return to Scotland, argues that

> if some people are disadvantaged by our
> willingness to allow the proximity of wolves
> back into our lives, people like sheep farmers
> and hunters, then that is simply part of the price
> that we pay for the privilege of a closer walk
> with natural forces, part of the debt that we owe
> for all that we have taken out of nature for far
> too long.[126]

But although the fate of our natural landscapes is a problem that belongs to us all, is it fair, practical, or moral for the debt we all owe to be paid by so few?

Do Wolves Belong?

Some contend that wolves simply have no place here. That the landscape has changed too much since wolves last padded across it. That wolves do not belong in the cultural landscapes that make up Britain and Ireland, but in the 'true wildernesses' of North America and Siberia.[127]

But this line of thinking is intertwined with cultural ideas of natural history. The mythical Caledonian Forest thought to have enswathed much of Scotland is often said to have been the haunt of wolves, perpetuating a notion that wolves require vast tracts of unbroken woodland.[128] Yet, although wolves are frequently conceptually linked with woodland, they do not need such places to survive. Arctic wolves manage without any trees at all. In Britain and Ireland, woodland was already in steep decline by the medieval period, and much depleted by the time wolves became extinct. What tree cover remained did provide wolves with shelter from persecution in the lead-up to their extinction, but it was not ecological need that led them to such places. While woodland remains an important place of shelter for wolves in our anthropocentric world, especially for raising their pups, the lack of forest cover in Britain and Ireland is not necessarily a barrier to reintroduction. Potential ungulate prey in the Scottish Highlands live largely in heath and upland bog, in areas not densely populated by people. So long as they could find places where they would be relatively undisturbed, especially to raise their pups, these habitats would be more suited to wolves than they may appear.[129]

Unlike the history of sheep and deer, the history of wolves in Britain and Ireland is not writ large across the landscape. It is obscured to the point that one Reddit user, prompted by the abundance of 'myths surrounding them' but apparent lack of 'actual evidence' of their existence, asked: 'did wolves actually live in the UK?'[130] For others, wolves are the denizens of a distant, frightful, wild, and uncivilised past, an animal now unimaginable and out of place in our current landscapes. As children we are fed stories of 'Old MacDonald', of jolly, rotund farmers and their happy animals,[131] in contrast to the Big Bad Wolves of our fairy tales and nightmares. From this early age we are taught that wolves are undesirable monsters and that farming is the beating heart of the countryside, the alternative a frightful state of 'natural anarchy in which every kind of beast roams our green and pleasant land, ripping other animals, and possibly humans, to shreds', as one journalist put it.[132] It matters little that wolves were found here as recently as the seventeenth century, nor that they were here, shaping our ecosystems, for far longer than humans were, or that they lived alongside people for thousands of years after that.[133] For many, they belong here about as much as the hyaenas and lions that disappeared tens of thousands of years ago. But of all the animals that have disappeared from Britain and Ireland, wolves are among some of the most recent inhabitants. The landscape has changed less in the intervening years than it has since other species disappeared,[134] yet some animals that became extinct much longer ago, like beavers and cranes, have been successfully restored.

It is certainly true that there are very few landscapes in Britain and Ireland that have not been altered by people. Much of the land is now managed in some capacity by a pervasive human population that has only increased since the last wolf trod these shores. But there is plenty of space for both humans and wolves. Wolves are highly adaptable, capable of living in almost any type of landscape,[135] including those which are heavily used by people. Although wolves are often associated with the open wildernesses of North America, there are in fact twice as many wolves in the half-the-size yet twice-as-densely-populated European continent,[136] where they occupy landscapes lived and worked in by humans.[137] Wolf populations are capable of recovering in places where the human population density is as high as 142 people per square kilometre.[138] It is not human presence in itself that affects their ability to survive, but the landscape of human hatred, fear, and intolerance.

If wolves truly needed wilderness, they simply would not survive in Europe.[139] Protected areas such as national parks undoubtedly do provide 'core refuges' for animals such as wolves, but viable populations could not be sustained solely within most these areas. Europe contains only a handful of protected areas of more than 1,000 square kilometres, and so most European wolves live in unprotected regions.[140] Although keeping wolves within places where there is less overlap with human land use gives them the best chance of survival by avoiding conflict,[141] it is more important to grant large carnivores space within wider landscapes that contain people, rather than stringently attempting to maintain a divide between nature and culture by confining wolves to small pockets of protected 'wilderness' where there are no people at all.[142]

One particular area of Britain is well placed to welcome back wolves: the Scottish Highlands.[143] Studies have suggested that an area of between 10,139km^2 and 18,857km^2 spanning the Scottish Highlands and Grampians could support between 200 and 376 wolves,[144] more than the suggested minimum viable population of 200 to 250 animals.[145] Although there are no continuous protected areas that could house a wolf population, which would instead have to occupy a mixture of private and public land,[146] the Cairngorms National Park alone is more than 4,500 square kilometres in size, about half as large as Yellowstone National Park,[147] where around 124 wolves lived at the end of 2023 (though their numbers have reached as many as 174 in previous years). The Highlands is also one of the least densely populated areas in Europe,[148] far below the maximum density at which it is estimated that wolves can recover, with just nine people per square kilometre. The road density, another measure of habitat suitability for wolves, is also low.[149] Although much of the landscape is used for sheep farming, livestock densities are relatively low and numbers of sheep have fallen since the 1990s.[150] And there is of course plenty of prey in the form of deer, as well as abundant upland and woodland areas for wolves to retreat to.

Reintroduction to Ireland is trickier. Although the island is not densely populated except for within its urban areas, much of rural Ireland is used for livestock farming. Combined with the fact that Ireland's national parks and other protected areas are not extensive, it has been suggested that there is not enough space where wolves could live relatively undisturbed and in sufficient numbers for a population to survive.[151] However, modelling has suggested that very small 'metapopulations' of wolves could survive in four protected areas: Wicklow Mountains National Park, Glenveagh National Park (Donegal), Killarney National Park (Kerry), and Wild Nephin National Park (Mayo). Though small, at 230km^2, 154.84km^2, 102.9km^2, and 150km^2 respectively, each park could support a single wolf pack.[152] While these would not be viable self-sustaining populations, they could be used for localised ecosystem recovery rather than species restoration, an approach that is becoming increasingly common.[153] Management would be tricky, however, especially given the need to keep an expanding population within the limited confines of isolated protected areas. And would such a heavily managed population constitute a true return of the wolf to Ireland? Although most mammals across Britain and Ireland today are managed to varying degrees, about whose 'wildness' questions are rarely raised, many may disagree with such hands-on management when it comes to an animal thought to be emblematic of wilderness itself.

In any case, just because we designate a particular space for wolves does not mean that they will stay there. Wolves are by nature roaming creatures, able to travel many miles over the course of a single day. While pack territories average between 100 and 200km^2, the high and low ends of the spectrum are extreme. Territories as small as 33km^2 and as vast as 6,272km^2 have been recorded.[154] Young wolves also frequently leave their natal packs to find territories of their own, journeys that may take them beyond the areas designated for them and close to towns and cities. On occasion, such journeys have been several hundred kilometres long.

That being said, territory sizes tend to be determined largely by prey availability. One study in the United States found that wolf territory sizes were just 116km^2 at deer densities of 6.2 per km^2, a figure half that of the average density of red deer in the Highlands.[155] It therefore seems unlikely that the territories of wolves reintroduced to the Scottish Highlands would be particularly large, nor that dispersing wolves would have to travel far to find suitable places to start their own packs. Even if wolves did venture further than expected, there would be protocols in place to prevent them from wandering too far, or into places they would not be welcome.

Various methods can also be employed to encourage reintroduced populations to remain in particular areas. Wolves can be kept in acclimation pens within the intended 'wolf area' before they are allowed to disperse into the wider landscape, a technique known as a 'soft release' which can encourage them to carve out territories in the vicinity rather than venturing too far once the gates are opened.[156] Wolves can also be fitted with electric collars that emit a shock when they approach a buried wire, a technique teaching them to avoid boundaries that have been set for them. The wolf's natural territoriality can also be capitalised upon by using wolf scents to simulate the boundaries of a neighbouring pack's territory.[157]

Another option is experimentally reintroducing wolves to large 'fenced reserves', a long-held ambition of Sutherland estate owner Paul Lister. There are successful precedents for a model of this kind in South Africa, where a 'managed metapopulation' of painted wolves lives in numerous fenced reserves, between which members are translocated in order to mimic natural dispersal.[158] Several studies have indicated the ecological viability of this approach in a Scottish context, suggesting that wolves would be far more effective at controlling deer numbers in such areas since they could reach higher densities, thus allowing them to exert greater pressure on their prey.[159]

Reintroducing wolves to enclosed areas would also offer unparalleled research opportunities to monitor their ecological impacts. This would be internationally significant research, which could help answer some of the biggest questions surrounding rewilding with large carnivores and the 'landscape of fear'.[160] The economic impacts could also be monitored to discover whether, as is predicted by researchers, the financial benefits of wolves might outweigh the management costs.[161] If wolves proved socioeconomically and ecologically beneficial within a fenced context, they may be more welcome in unfenced areas in the future.

But though this approach would have its benefits, it also raises significant questions. Is a managed wolf the same as a wolf that roams free, and is a reintroduction with fences a true reintroduction? Animal ethicist William S. Lynn argues that wolves kept entirely separate from humans constitute simply 'representative samples of biodiversity' and 'nothing more than another commodity of human society […] relegated to ghettos called refuge and wilderness'. For Lynn, true wolf recovery means not just the return of a healthy wolf population, but specifically that the population constitutes 'individuals [who] are respected as co-residents in wild and humanized landscapes'.[162]

There are also ethical and moral questions about fencing in an animal which, by nature, is wide-ranging and highly territorial. Dispersing wolves with nowhere to go could be at risk of fatal conflicts with existing packs. There is also the question of the welfare of prey species. Animal welfare laws prohibit housing live prey animals in the same enclosure as their predators, which would require either fencing that would somehow keep wolves firmly within the reserve while allowing prey species to exit, or exceptions to or reinterpretations of the law to distinguish 'wilderness reserve' from 'enclosure'.[163] Fencing could also have dangerous implications for Scotland's fiercely protected Right to Roam, potentially offering landowners a loophole to exclude people from their land unless they paid for the privilege. In such cases, wolves would become mere tokens of biodiversity – and highly divisive ones at that.

An alternative suggestion involves reintroducing wolves to the Isle of Rùm, a largely uninhabited island off the west coast of Scotland. At just over 100km^2, most of Rùm's mountainous landscape is managed as a nature reserve. A similar reintroduction was successfully carried out on Isle Royale, a lake island in the state of Michigan, after the local wolf population reached unsustainably low levels and moose numbers exploded. Given its high red deer density, the Isle of Rùm could potentially support an estimated 20 or so wolves. Like a fenced-reserve reintroduction, a project like this could offer valuable insights into the potential ecological outcomes of a larger-scale reintroduction project on the mainland.[164]

Reintroductions must also go hand in hand with sound management, raising questions of how wolves would be managed, who by, and how much intervention would be considered acceptable. For some, a wolf population that is heavily managed, monitored, and manipulated would not be 'natural', raising questions over the point of a reintroduction that ultimately maintained an 'artificial' population. For others, 'naturalness' may be a sacrifice that we – and the wolves – would have to make. These days, very few landscapes and animal populations in Britain and Ireland are entirely unmanaged. Like it or not, wolves could not just be left to their own devices. We have to be willing to manage wolves, whatever that looks like (even if it involved killing them), based on sound ecological reasoning in balance with social needs.

Ultimately, even if every last detail was considered, planned, and honed, there is no seeing into the future. Things go wrong. Animals are unpredictable, people perhaps more so.

What happens then?

POSTSCRIPT

The Future of Wolf Land

We are disconnected from our wild past and heading towards an uncertain future. In a world of collapsing ecosystems and a volatile climate, we must repair the broken bridges between our species and the rest of the natural world. We must reconnect with nature not as the inchoate 'out there' where wilderness seeps like sludge, threatening to swallow centuries of human civilisation, but as part of ourselves and we part of it. We must find compromise between our own needs and desires and the needs of the planet which sustains us – a world that we need far more than it needs us.

Part of that compromise must be to carve out space for animals that we find difficult to accommodate. We will have to learn to prioritise species other than our own, even – and perhaps especially – those we least want to, those with which we have historically struggled and which we continue to struggle to live alongside.

As island nations, Britain and Ireland can pick and choose which animals to welcome and those that we would rather relegate to the history books. So far, we have used that power to keep far more animals out than we have let in. As a result, ours are some of the most starkly nature-depleted countries on the planet.[1] According to the Biodiversity Intactness Index (which ranks each country assessed according to the abundance of native species still present), Britain and Ireland are among some of the least biodiverse countries in the world, both ranking in the bottom 15 of the 240 countries that have been assessed. A healthy ecosystem has at least 90% of native biodiversity intact,[2] but Britain and Ireland sit at just 42% and 40% respectively.[3] These figures are alarmingly close to the 30% at which an ecosystem is considered so depleted that it is non-functional.[4]

We should ask not only 'What if we brought wolves back?', but also 'What if we did not?'. What is the natural future of the Britain and Ireland without the wolf? What is the alternative? A future that contains wolves is far more certain than a future in which their ghosts haunt our increasingly fuzzy natural memories as we desperately attempt to protect a world that's already ablaze.

To forge a new way of living in a wolfish world will undoubtedly be difficult. For hundreds of years we have been the masters of the landscape – it will not be easy to suddenly relinquish even a small part of that control. Learning to live with wolves once more will take a shift in our worldviews, our practices, our cultures, our selves. It will take overcoming a fear that has been thousands of years in the making.

But wolves have not always been hated and feared, nor are they hated and feared everywhere today. Lupophobia is not the natural state of humankind.

In Britain and Ireland today, our lives and our relationship with nature are shaped by the absence of predators. Our natural identity is intricately tied up with sheep farming, with patchworks of fields in every shade of brown, green and yellow, and with bluebell-carpeted woodlands and with quaint trickling rivers. We no longer know how to walk through a wolf's wood or a lynx's landscape. We are a nation of foxes, hedgehogs, and sparrows, creatures that live and feed in our gardens. Our 'wildernesses' house ungulates in numbers so great that they resemble fields full of livestock. We are the inheritors of a land created by the absence of wolves, built on their bones and blood.

But wolves are a part of our heritage, our culture, our natural world, and our collective subconsciousness. We shared these islands with them for thousands of years, learning from them, emblematising them, guarding our flocks against them, hunting alongside them, finding ways to coexist with them. We used their pelts to stay warm, their companionship to give us comfort, their predatorial prowess to help us hunt our prey, and their body-parts for medicine and spiritual protection. Even after the Agricultural Revolution, we still managed to coexist with wolves, both species adapting to one another's presence. When our wolf population died out, so too did old ways of life, old relationships, old connections to the land. We lost far more than we gained when we eradicated our wolves.

They were there throughout every stage of history. They roamed a frozen world alongside bears, hyaenas, lions, bison, and mammoths.

They saw the coming of humans and, later, the placid animals that they cultivated. They saw members of their own species forge a strange, symbiotic relationship with people until they were no longer recognisable as wolves. They witnessed the rise and fall of civilizations: the arrival of the Celts, the coming and going of the Romans, the invasions of Germanic tribes, of the Danes, of the Normans. They survived through it all until humankind became too many, too widespread, too intent on removing them from their homes. Wolves challenged the authority of humans, their dominion, and their civilising aspirations like no other animal, and were accordingly transformed from simply another creature with which we shared the landscape into competitors, vermin, thieves, intruders, and trespassers. Until, finally, all that was left were perverse monuments to the slaying of 'last wolves', celebrations of the final blow in a sadistic and ecologically disastrous murder campaign.

From the Ice Age to the medieval period, from the Renaissance to the present day, wolves have always occupied our minds. They have served as vectors and vehicles of our own thoughts and ideas about ourselves, our relationship to the natural world, and about life itself. We wove them into the fabric of our stories, inextricably entangling them into our cultural lives. We are connected to them emotionally, culturally, and psychologically, and our lives are richer for it. Our lives would be richer still if we learned to live alongside wolves not just on the page or screen, but in the real world too.

We tend to think of ourselves as forward thinking and superior, to consider the past as dim, distant, and dark, and the humans that inhabited it as unintelligent, unsophisticated, perhaps even savage. For some, the eradication of wolves may seem a sign of increased intelligence, a step out of the 'Dark Ages' towards civilisation and a tamed world. As Freud wrote in his *Civilisation and its Discontents*:

> a country has a high level of civilization if we find that in it everything that can assist man in his exploitation of the land and protect him against the forces of nature – everything, in short, that is of use to him – is attended to and properly ordered. In such countries, rivers that threaten to flood the land must have their courses regulated and their waters channelled to areas of drought. The soil

> must be carefully tilled and planted with crops that it is suited to support [...]. Dangerous wild beasts must be exterminated, and the breeding of domestic animals must flourish.[5]

In his *Being a Beast*, writer Charles Foster even classes 'modernity' as beginning when wolves disappeared.[6]

But in taking these steps towards civilisation we have made ourselves vulnerable, digging our own graves by destroying the natural world and the millions of other species that make up the web of life upon which we depend. As Jack London once wrote, 'civilization [...] has spread a veneer over the surface of the soft-shelled animal known as man. It is a very thin veneer; but so wonderfully is man constituted that he squirms on his bit of achievement and believes he is garbed in armor-plate'.[7] But we are only powerful so long as there is something for us to have power over. Soon, we will be the rulers of an empty, dying world.

Those who killed wolves and so many other species in centuries past knew little of the damage they were causing, the havoc they would unleash upon the environment. We have no such excuse today, and yet we continue to burn, pollute, waste, and kill, to put ourselves and our own interests above all other species, even our own future generations who will inherit our degraded world, who won't forgive us if we fail to act.

It is us, today, who are 'backwards', unwilling to accept the importance of the natural world which sustains us even as we head towards climatic doom and ecosystem collapse. If we cannot use our modern technology, our scientific advancements, our increased intelligence, our awareness of the importance of the ecosystems around us, and our moral and emotional sophistication to live alongside wolves and other predators, then are we so clever, so advanced, so 'civilised' after all?

The history of the wolf in Britain and Ireland is far from complete. There is more of the story left to tell, more manuscripts to be pored over, more remains to be unearthed, more lost tomes waiting to be rediscovered, more scattered references to be brought to light.

And perhaps the story is not yet over.

Perhaps, in a hundred years' time, someone will write a new history of wolves in Britain and Ireland that includes a new chapter, one detailing their long-overdue return to their old haunts and how they breathed new life into the landscape and the people who lived in it. Perhaps the wolf of ink and paper and fear and fallacy will be decisively disentangled from the wolf of flesh and blood. A wolf who breathes the cold air of these isles and, as her howls drift over heathland, mountains, and forests, makes Wolf Land come alive once more.

Notes and References

Introduction: The Wolf of Time and (Hi)story

1. Margaret Atwood, *The Blind Assassin* (London: Virago Press, 2011), p. 423.
2. Cited in Mary Colwell, *Beak, Tooth and Claw: Living with Predators in Britain* (London: Collins, 2021), p. 29.
3. Barry Holstun Lopez, *Of Wolves and Men* (New York, NY: Charles Scribner's Sons, 1978), p. 4.
4. L. David Mech and Luigi Boitani, 'Wolf Social Ecology', in *Wolves: Behavior, Ecology, and Conservation*, ed. by L. David Mech and Luigi Boitani (Chicago, IL: University of Chicago Press, 2003), pp. 1–34 (p. 6). See also Garry Marvin, *Wolf* (London: Reaktion Books, 2012), pp. 33–4.
5. As anthropologist Garry Marvin notes, the true history of the wolf is inaccessible to us. We can only write a history of what humans have thought about wolves and how they have interacted with wolves in both the landscape and in culture. Even archaeological evidence, which may seem to have an air of objectivity, is interpreted and made sense of by people; 'Wolves in Sheep's (and Others') Clothing', in *Beastly Natures: Animals, Humans, and the Study of History*, ed. by Dorothee Brantz (Charlottesville: University of Virginia Press, 2010), pp. 59–78.
6. Yan Ge, *Strange Beasts of China*, trans. by Jeremy Tiang (London: Tilted Axis Press, 2020), p. 45.
7. James Edmund Harting, *British Animals Extinct Within Historic Times* (Boston: Osgood and Co., 1880), p. 165.
8. Mech and Boitani, 'Wolf Social Ecology', pp. 1–3.
9. Marvin, *Wolf*, p. 19.
10. Marvin, *Wolf*, pp. 16–18.
11. L. David Mech, *The Wolf: The Ecology and Behaviour of an Endangered Species* (Minneapolis, MN: University of Minnesota Press, 1970), p. 246.
12. Rolf O. Peterson and Paolo Ciucci, 'The Wolf as a Carnivore', in *Wolves: Behavior, Ecology, and Conservation*, ed. by Mech and Boitani, pp. 104–30 (p. 119).
13. Robert H. Busch, *The Wolf Almanac: A Celebration of Wolves and Their World*, new and rev. edn (Guilford, CT: Lyons Press, 2007), p. 46, and Luigi Boitani, *Action Plan for the Conservation of the Wolves (*Canis lupus*) in Europe* (Strasbourg: Council of Europe Publishing, 2000), p. 19.
14. L. David Mech and Rolf O. Peterson, 'Wolf-Prey Relations', in *Wolves: Behavior, Ecology, and Conservation*, ed. by Mech and Boitani, pp. 131–60 (pp. 144–5).
15. Douglas W. Smith and Gary Ferguson, *Decade of the Wolf: Returning the Wild to Yellowstone*, rev. edn (Guilford, CT: Lyons Press, 2012), pp. 119–20.
16. Boitani, *Action Plan for the Conservation of the Wolves in Europe*, p. 16.
17. Marvin, *Wolf*, pp. 13–14.
18. L. David Mech, '*Canis lupus*', *Mammalian Species*, 37 (1974), 1–6 (p. 5).
19. L. David Mech, 'The Scientific Classification of Wolves: *Canis lupus soupus*', *International Wolf*, 21.1 (2011), 4–7 (pp. 5–6).

20. Kieran Hickey, *Wolves in Ireland: A Natural and Cultural History* (Dublin: Four Courts Press, 2011), pp. 108–9.
21. This history is organised into seven chronological chapters, within which the material is organised largely chronologically, though sometimes thematically.

Chapter 1: Wolves in a Frozen Land

1. Xiaoming Wang and Richard H. Tedford, *Dogs: Their Fossil Relatives and Evolutionary History* (New York, NY: Columbia University Press, 2010), p. 51.
2. Ibid., p. 56.
3. Ibid., p. 148.
4. Ibid., and Lucy O. H. Flower and Danielle C. Schreve, 'An Investigation of Palaeodietary Variability in European Pleistocene Canids', *Quaternary Science Reviews*, 96 (2014), 188–203 (p. 200).
5. Flower and Schreve, 'Palaeodietary Variability in European Pleistocene Canids', p. 189 (Figure 1), and Wang and Tedford, *Dogs*, p. 148. For more on early wolf evolution, see M. Sotnikova and L. Rook, 'Dispersal of the Canini (Mammalia, Canidae: Caninae) Across Eurasia During the Late Miocene to Early Pleistocene', *Quaternary International*, 212 (2010), 86–97, and Ronald M. Nowak, 'Wolf Evolution and Taxonomy', in *Wolves: Behavior, Ecology, and Conservation*, ed. by Mech and Boitani, pp. 239–58.
6. Ralf-Dietrich Kahlke et al., 'Western Palaearctic Palaeoenvironmental Conditions During the Early and Early Middle Pleistocene Inferred from Large Mammal Communities, and Implications for Hominin Dispersal in Europe', *Quaternary Science Reviews*, 30 (2011), 1368–95 (p. 1386).
7. Ádám Miklósi, *Dog Behaviour, Evolution, and Cognition* (Oxford: Oxford University Press, 2015), p. 68.
8. See Lorenzo Rook, 'The Plio-Pleistocene Old World *Canis* (*Xenocyon*) ex gr. *falconeri*', *Bolletino della Società Paleontologica Italiana*, 33 (1994), 71–82, and Wang and Tedford, *Dogs*, pp. 149–50.
9. Some have argued that *Canis arnensis* is more likely to be the ancestor of the Mosbach wolf (Flower and Schreve, 'Palaeodietary Variability in European Pleistocene Canids', p. 190), an issue complicated by the fact that despite being more jackal-like overall, *Canis arnensis* has a skull which displays more similarities to *Canis lupus* than that of *Canis etruscus*; see Marco Cherin et al., 'Re-Defining *Canis etruscus* (Canidae, Mammalia): A New Look into the Evolutionary History of Early Pleistocene Dogs Resulting from the Outstanding Fossil Record from Pantalla (Italy)', *Journal of Mammalian Evolution*, 21 (2014), 95–110 (p. 107).

    It has also been proposed that a newly discovered species recently unearthed in Czechia, *Canis lupus bohemica*, is the ancestor of *Canis mosbachensis*; Cajus G. Dietrich, 'Eurasian Grey and White Wolf Ancestors—800,000 Years Evolution, Adaptation, Pathologies and European Dog Origins', *Acta Zoologica*, 105 (2022), 2–37.
10. Sotnikova and Rook, p. 95, and L. O. H. Flower, 'New Body Mass Estimates of British Pleistocene Wolves: Palaeoenvironmental Implications and Competitive Interactions', *Quaternary Science Reviews*, 149 (2016), 230–47 (p. 231).
11. This is the approximate date of the oldest *Canis mosbachensis* remains found in Britain, which were unearthed in Norfolk; Flower, 'New Body Mass Estimates of British Pleistocene Wolves', p. 231.
12. Although the Pleistocene has ended, Earth is still in an 'ice age' since permanent ice sheets and glaciers persist today. Our current warm climate is simply the result of another interglacial period, although humans have now so profoundly

affected the climate that we have likely disrupted the natural cycles of glaciation and interglacials.

13. At the beginning of the Ice Age, Britain was permanently connected to the European mainland via a land bridge sitting atop the waves of what is now the Strait of Dover. This ridge, known as an isthmus, was initially breached by a flood around 450,000 to 425,000 years ago, which caused extensive erosion but did not completely destroy it. Around 160,000 years ago there was a second flood, resulting in further erosion which ultimately left Britain completely isolated during subsequent periods when temperatures warmed, glaciers melted, and sea levels rose, submerging the ridge; Sanjeev Gupta et al., 'Two-stage Opening of the Dover Strait and the Origin of Island Britain', *Nature Communications*, 8:15101 (2017), and Philip Gibbard, 'Europe Cut Adrift', *Nature*, 448 (2007), 259–60. However, periods of glaciation saw sea levels dramatically lower as water was locked up in ice, revealing other paths between Britain and the Continent (most notably Doggerland, a large expanse of land now submerged beneath the North Sea) as well as land beyond the current coastline. Ice sheets also served as bridges between Britain and northern Europe.

    Ireland appears to have split from Britain by the time the isthmus between Britain and Europe was first breached (Alan Turner, 'The Evolution of the Guild of Large Carnivora of the British Isles During the Middle and Late Pleistocene', *Journal of Quaternary Science*, 24 (2009), 991–1005 (p. 993)), though the island was connected to Britain via an ice sheet during periods of glaciation. There is also debate as to whether sea levels were ever low enough during late-Ice Age glaciation to reveal land bridges between Ireland and Britain; R. J. Devoy, 'The Problem of a Late Quaternary Landbridge Between Britain and Ireland', *Quaternary Science Reviews*, 4.1 (1985), 43–58.

14. Turner, 'The Evolution of the Guild of Large Carnivora', p. 999.
15. Very few sites in Britain have yielded remains of the elusive Mosbach wolf, the majority of which date to the Cromerian Interglacial. The species is unknown in Ireland, and there is also currently no evidence to suggest that *Canis mosbachensis* was present in Britain during periods of glaciation, perhaps because land connecting Britain to mainland Europe allowed them to move further south during colder periods; Flower, 'New Body Mass Estimates of British Pleistocene Wolves', p. 239.
16. Rook, 'The Plio-Pleistocene Old World *Canis* (*Xenocyon*)', p. 72.
17. The Boxgrove Project <https://boxgroveproject.wordpress.com/about/> [accessed 25 November 2021].
18. Matthew Pope, 'Placing Boxgrove in Its Prehistoric Landscape', *Archaeology International*, 7 (2003), 13–16 (p. 14).
19. Flower, 'New Body Mass Estimates of British Pleistocene Wolves', pp. 238 and 240.
20. Michael J. Bishop, *The Mammal Fauna of the Early Middle Pleistocene Cavern Infill Site of Westbury-Sub-Mendip, Somerset* (London: The Palaeontological Association, 1982), p. 63, and Flower, 'New Body Mass Estimates of British Pleistocene Wolves', p. 239.
21. See Flower and Schreve, 'Palaeodietary Variability in European Pleistocene Canids', pp. 200–1; Flower, 'New Body Mass Estimates of British Pleistocene Wolves', p. 239; Carlo Meloro, 'Feeding Habits of Plio-Pleistocene Large Carnivores as Revealed by Themandibular Geometry', *Journal of Vertebrate Paleontology*, 31 (2011), pp. 428–46 (p. 437); and Laura Domingo et al., 'New Insights into the Middle Pleistocene Paleoecology and Paleoenvironment of the Northern Iberian Peninsula (Punta Lucero Quarry site, Biscay): A Combined Approach Using Mammalian Stable Isotope Analysis and Trophic Resource Availability Modeling', *Quaternary Science Reviews*, 169 (2017), 243–62 (p. 255).

22. Flower, 'New Body Mass Estimates of British Pleistocene Wolves', pp. 239–40.
23. Modern-day African painted wolves tend to avoid areas where lions are active, and for good reason: lions in South Africa have been known to prey on painted wolves; Flower, 'New Body Mass Estimates of British Pleistocene Wolves', p. 240.
24. Flower, 'New Body Mass Estimates of British Pleistocene Wolves', p. 240.
25. Bishop, *Mammal Fauna of the Early Middle Pleistocene Cavern Infill Site of Westbury-Sub-Mendip*, p. 63, and Flower, 'New Body Mass Estimates of British Pleistocene Wolves', pp. 239–40.
26. Flower, 'New Body Mass Estimates of British Pleistocene Wolves', p. 240.
27. There is continuing debate as to whether *Homotherium latidens*, the sabre-toothed cat found in Britain at this time, was solitary or social; Roman Croitor and Jean-Philippe Brugal, 'Ecological and Evolutionary Dynamics of the Carnivore Community in Europe During the Last 3 Million Years', *Quaternary International*, 212.2 (2010), 98–108 (p. 105), and Ross Barnett et al., 'Genomic Adaptations and Evolutionary History of the Extinct Scimitar-Toothed Cat, *Homotherium latidens*', *Current Biology*, 30.24 (2020), 5018–25 (pp. 5022–3).
28. Flower, 'New Body Mass Estimates of British Pleistocene Wolves', p. 240.
29. Turner, 'The Evolution of the Guild of Large Carnivora', p. 1000, and Saverio Bartolini-Lucenti et al., 'The Early Hunting Dog from Dmanisi with Comments on the Social Behaviour in Canidae and Hominins', *Scientific Reports*, 11 (2021), 1–10 (p. 3 (Figure 1a)).
30. The precise family tree is the subject of some debate, and it is not universally accepted that *Canis mosbachensis* is the immediate ancestor of *Canis lupus*. Various potential 'intermediate' species between *Canis etruscus*, *Canis mosbachensis*, and *Canis lupus* continue to be unearthed.
31. Richard H. Tedford, Xiaoming Wang, and Beryl E. Taylor, 'Phylogenetic Systematics of the North American Fossil Caninae (Carnivora: Canidae)', *Bulletin of the American Museum of Natural History*, 325 (2009), 1–218 (p. 181).
32. Dawid A. Iurino et al., 'A Middle Pleistocene Wolf from Central Italy Provides Insights on the First Occurrence of *Canis lupus* in Europe', *Scientific Reports*, 12 (2022), 1–13 (p. 9).
33. Liisa Loog et al., 'Ancient DNA Suggests Modern Wolves Trace their Origin to a Late Pleistocene Expansion from Beringia', *Molecular Ecology*, 29 (2020), 1596–1610 (p. 1606).
34. Flower, 'New Body Mass Estimates of British Pleistocene Wolves', p. 232, and Andrew P. Currant and Anne Eastham, 'The Fauna', in *Neanderthals in Wales: Pontnewydd and the Elwy Valley Caves*, ed. by Stephen Aldhouse-Green, Rick Peterson, and Elizabeth A. Walker (Oxford: Oxbow Books, 2012), pp. 100–17 (pp. 102 (Table 8.1) and 106).
35. Turner, 'The Evolution of the Guild of Large Carnivora', pp. 995 (Table 3 column 7) and 1000, and Flower and Schreve, 'Palaeodietary Variability in European Pleistocene Canids', p. 190.
36. Flower, 'New Body Mass Estimates of British Pleistocene Wolves', p. 241.
37. Ibid.
38. Flower and Schreve, 'Palaeodietary Variability in European Pleistocene Canids', p. 198, and Flower, 'New Body Mass Estimates of British Pleistocene Wolves', p. 241.
39. Flower, 'New Body Mass Estimates of British Pleistocene Wolves', pp. 241–2, and Flower and Schreve, 'Palaeodietary Variability in European Pleistocene Canids', pp. 198–9.
40. Flower, 'New Body Mass Estimates of British Pleistocene Wolves', p. 242.
41. Ibid., pp. 241–2.

42. Nicholas Barton, 'The Lateglacial or Latest Palaeolithic Occupation of Britain', in *The Archaeology of Britain: An Introduction from Earliest Times to the Twenty-First Century*, ed. by John Hunter and Ian Ralston, second edn (Abingdon and New York: Routledge, 2009), pp. 18–52 (p. 23), and Rebecca Wragg Sykes, 'Neanderthals in Britain: Late Mousterian Archaeology in Landscape Context' (unpublished doctoral thesis, University of Sheffield, 2009), pp. 73 and 94.
43. Geoff M. Smith, 'Neanderthal Megafaunal Exploitation in Western Europe and Its Dietary Implications: A Contextual Reassessment of La Cotte de St Brelade (Jersey)', *Journal of Human Evolution*, 78 (2015), 181–201 (p. 190).
44. Shumon T. Hussain, Marcel Weiss, and Trine Kellberg Nielsen, 'Being-with Other Predators: Cultural Negotiations of Neanderthal-carnivore Relationships in Late Pleistocene Europe', *Journal of Anthropological Archaeology*, 66 (2022), 1–37.
45. Andrew P. Currant and Roger Jacobi, 'The Mammal Faunas of the British Late Pleistocene', in *The Ancient Human Occupation of Britain*, ed. by Nick Ashton, Simon G. Lewis, and Chris Stringer (Amsterdam: Elsevier, 2011), pp. 165–80 (pp. 165–6), and Lucy Olivia Holman Flower, 'Canid Evolution and Palaeoecology in the Pleistocene of Western Europe, with Particular Reference to the Wolf *Canis lupus* L. 1758' (unpublished doctoral thesis, Royal Holloway University of London, 2014), p. 547.
46. Flower, 'Canid Evolution', p. 334.
47. I. De Groote, M. Lewis, and C. Stringer, 'Prehistory of the British Isles: A Tale of Coming and Going', *Bulletins et mémoires de la Société d'anthropologie de Paris*, 30 (2017), 1–13 (p. 8).
48. Flower, 'Canid Evolution', pp. 382–3, and Turner, 'The Evolution of the Guild of Large Carnivora', p. 994 (Table 2 column 5e).
49. Turner, 'The Evolution of the Guild of Large Carnivora', p. 995 (Table 3 column 5e).
50. Remains of these wolves have 'low percentages of heavily worn teeth', suggesting that they were not often forced to eat bone; Flower, 'Canid Evolution', p. 334.
51. Currant and Jacobi, 'The Mammal Faunas of the British Late Pleistocene', pp. 167 and 168 (Table 10.2); Flower, 'Canid Evolution', p. 383; and Turner, 'The Evolution of the Guild of Large Carnivora', pp. 994–5 (Table 2 column 5c, and Table 3 column 5c).
52. Flower and Schreve, 'Palaeodietary Variability in European Pleistocene Canids', p. 190.
53. Flower, 'New Body Mass Estimates of British Pleistocene Wolves', p. 242. That said, it has been suggested that small relict populations of these predators may have survived in the north; Turner, 'The Evolution of the Guild of Large Carnivora', p. 1000.
54. Flower and Schreve, 'Palaeodietary Variability in European Pleistocene Canids', p. 190.
55. Flower, 'Canid Evolution', pp. 461–2 and 384, and Currant and Jacobi, 'The Mammal Faunas of the British Late Pleistocene', p. 169 (Table 10.3).
56. Flower and Schreve, 'Palaeodietary Variability in European Pleistocene Canids', p. 198, and Flower, 'New Body Mass Estimates of British Pleistocene Wolves', p. 242.
57. Flower and Schreve, 'Palaeodietary Variability in European Pleistocene Canids', p. 198.
58. Flower, 'New Body Mass Estimates of British Pleistocene Wolves', p. 242.
59. Turner, 'The Evolution of the Guild of Large Carnivora', p. 1001.
60. Flower and Schreve, 'Palaeodietary Variability in European Pleistocene Canids', pp. 197–8.

61. Flower, 'Canid Evolution', pp. 493–4; Flower and Schreve, 'Palaeodietary Variability in European Pleistocene Canids', pp. 197–8; and Flower, 'New Body Mass Estimates of British Pleistocene Wolves', pp. 242 and 244.
62. Katharine Scott, 'The Large Vertebrates from Picken's Hole, Somerset', *Proceedings of the University of Bristol Spelæological Society*, 27 (2018), 267–313 (pp. 289–90).
63. Flower, 'New Body Mass Estimates of British Pleistocene Wolves', p. 243; Currant and Jacobi, 'The Mammal Faunas of the British Late Pleistocene', pp. 171–2; and Turner, 'The Evolution of the Guild of Large Carnivora', p. 1000.
64. Flower, 'New Body Mass Estimates of British Pleistocene Wolves', p. 243.
65. Humans made it to Ireland by at least 33,000 years ago, though evidence of permanent settlement is more recent (see pp. 37–8 below).
66. Flower, 'New Body Mass Estimates of British Pleistocene Wolves', pp. 243–4.
67. Roger Jacobi, Nick Debenham, and John Catt, 'A Collection of Early Upper Palaeolithic Artefacts from Beedings, near Pulborough, West Sussex, and the Context of Similar Finds from the British Isles', *Proceedings of the Prehistoric Society*, 73 (2007), 229–326 (p. 278).
68. Steven Mithen, 'The Hunter-Gatherer Prehistory of Human–Animal Interactions', *Anthrozoös*, 12 (1999), 195–204 (p. 200).
69. Danielle Caroline Schreve, 'Mammalian Biostratigraphy of the Later Middle Pleistocene in Britain' (unpublished doctoral thesis, University College London, 1998), pp. 610–11.
70. Flower, 'New Body Mass Estimates of British Pleistocene Wolves', pp. 242–4; Flower and Schreve, 'Palaeodietary Variability in European Pleistocene Canids', p. 198; and Turner, 'The Evolution of the Guild of Large Carnivora', p. 1002.
71. *Palaeontological Memoirs and Notes of the Late Hugh Falconer, with a Biographical Sketch of the Author*, ed. by Charles Murchison, 2 vols (London: Hardwicke, 1868), II: *Mastodon, Elephant, Rhinoceros, Ossiferous Caves, Primeval Man and His Contemporaries*, pp. 462–3.
72. Richard Lydekker, *Catalogue of the Fossil Mammalia in the British Museum*, 5 vols (London: Taylor and Francis, 1885–7), I: *The Orders Primates, Chiroptera, Insectivora, Carnivora, and Rodentia* (1885), p. 122.
73. Sidney H. Reynolds, *A Monograph of the British Pleistocene Mammalia, Vol. II, Part III: The Canidæ* (London: Printed for the Palaeontographical Society, 1909), p. 9.
74. P. David Polly and Jussi T. Eronen, 'Mammal Associations in the Pleistocene of Britain: Implications of Ecological Niche Modelling and a Method for Reconstructing Palaeoclimate', in *The Ancient Human Occupation of Britain*, ed. by Ashton, Lewis, and Stringer, pp. 279–304 (p. 298).
75. Terry O'Connor, 'Introduction – The British Fauna in a Changing World', in *Extinctions and Invasions: A Social History of British Fauna*, ed. by Terry O'Connor and Naomi Sykes (Oxford: Windgather Press, 2010), pp. 1–9 (p. 1).
76. Turner, 'The Evolution of the Guild of Large Carnivora', p. 1002.
77. Currant and Jacobi, 'The Mammal Faunas of the British Late Pleistocene', pp. 173–4, and Stephen Aldhouse-Green et al., 'The Nature of Human Activity at Cae Gwyn and Ffynnon Beuno Caves and the Dating of Prey and Predator Presences', in *No Stone Unturned: Papers in Honour of Roger Jacobi*, ed. by Nick Ashton and Claire Harris (London: Lithic Studies Society, 2015), pp. 77–92 (pp. 84–8).
78. Currant and Jacobi, 'The Mammal Faunas of the British Late Pleistocene', p. 173.
79. Lia Tarle, 'Clothing and the Replacement of Neanderthals by Modern Humans' (unpublished MA dissertation, Simon Fraser University, 2012), pp. 30–1.
80. Ibid., pp. 32, 22, and 104–5.

81. Flower, 'Canid Evolution', p. 386.
82. Andrew C. Kitchener, 'Extinctions, Introductions and Colonisations of Scottish Mammals and Birds Since the Last Ice Age', in *Species History in Scotland*, ed. by Robert A. Lambert (Edinburgh: Scottish Cultural Press, 1998), pp. 63–92 (p. 74).
83. Nicola Anne Murray, 'The Behavioural Ecology of Reindeer (*Rangifer tarandus*) During the Last Glaciation in Britain and Its Implications for Human Settlement, Subsistence and Mobility (unpublished doctoral thesis, University of Edinburgh, 1993), pp. 111–17, and Dougie Strang, 'Creag nan Uamh', *Bella Caledonia*, 3 November 2018 <https://bellacaledonia.org.uk/2018/11/03/creag-nan-uamh/> [accessed 30 October 2021].
84. Strang, 'Creag nan Uamh'.
85. Peter Woodman, *Ireland's First Settlers: Time and the Mesolithic* (Oxford: Oxbow Books, 2015), pp. 18–19, and W. Ian Montgomery et al., 'Origin of British and Irish Mammals: Disparate Post-glacial Colonisation and Species Introductions', *Quaternary Science Reviews*, 98 (2014), 144–65 (p. 156).
86. Montgomery et al., 'Origin of British and Irish Mammals', p. 156.
87. Hickey, *Wolves in Ireland*, p. 22.
88. G. R. Coope, 'Fossil Coleopteran Assemblages as Sensitive Indicators of Climatic Changes During the Devensian (Last) Cold Stage [and Discussion]', *Philosophical Transactions of the Royal Society B*, 280 (1977), 313–40 (p. 321).
89. Montgomery et al., 'Origin of British and Irish Mammals', p. 156.
90. Humans may have visited Ireland at this time, but they do not appear to have stuck around; J. P. Mallory, *The Origins of the Irish* (London: Thames & Hudson, 2013), p. 39.
91. Currant and Jacobi, 'The Mammal Faunas of the British Late Pleistocene', pp. 175–6, and A. P. Currant, 'The Lateglacial Mammal Fauna of Gough's Cave, Cheddar, Somerset', *Proceedings of the University of Bristol Spelæological Society*, 17 (1986), 286–304 (p. 293).
92. Currant, 'The Lateglacial Mammal Fauna of Gough's Cave', p. 298, and S. N. Collcutt, A. P. Currant, and C. J. Hawkes, 'A Further Report on the Excavations at Sun Hole, Cheddar', *Proceedings of the University of Bristol Spelæological Society*, 16 (1981), 21–38 (p. 33).
93. Collcutt, Currant, and Hawkes, pp. 33–4.
94. Roger Jacobi, 'The Late Upper Palaeolithic Lithic Collection from Gough's Cave, Cheddar, Somerset and Human Use of the Cave', *Proceedings of the Prehistoric Society*, 70 (2004), 1–92 (p. 78).
95. Roger Jacobi and Tom Higham, 'The Later Upper Palaeolithic Recolonisation of Britain: New Results from AMS Radiocarbon Dating', *Developments in Quaternary Sciences*, 14 (2011), 223–47 (pp. 230 and 233).
96. Christopher Smith, *Late Stone Age Hunters of the British Isles* (London and New York: Routledge, 1992), p. 98.
97. Currant and Jacobi, 'The Mammal Faunas of the British Late Pleistocene', p. 174.
98. Jacobi and Higham, 'The Later Upper Palaeolithic Recolonisation of Britain', pp. 233 and 242.
99. Gianpiero Di Maida et al., '*Lupus in Fabula*: The Representation of the Wolf (Canis lupus) in European Palaeolithic Art', in *Dogs, Past and Present: An Interdisciplinary Perspective*, ed. by Ivana Fiore and Francesca Lugli (Oxford: Archaeopress, 2023), pp. 312–18 (pp. 313–14).
100. Mithen, 'The Hunter—Gatherer Prehistory of Human—Animal Interactions', p. 199.
101. Di Maida et al., '*Lupus in Fabula*', p. 312, citing Claude Lévi-Strauss, *Le Totémisme aujourd'hui* (Paris: Presses Universitaires de France, 1962). For a

translation, see Claude Lévi-Strauss, *Totemism*, trans. by Rodney Needham (Boston, MA: Beacon Press, 1963), p. 89.

102. Di Maida et al., '*Lupus in Fabula*', p. 317.
103. Pat Shipman, *The Invaders: How Humans and Their Dogs Drove Neanderthals to Extinction* (Cambridge, MA: Belknap Press, 2015), pp. 210–11.
104. Di Maida et al., '*Lupus in Fabula*', p. 317.
105. Shipman, *The Invaders*, pp. 211–12.
106. Matthew Beresford, *The White Devil: The Werewolf in European Culture* (London: Reaktion Books, 2013), p. 20.
107. Ibid., p. 20, citing A. Leslie Armstrong, 'Pin Hole Cave Excavations, Creswell Crags, Derbyshire. Discovery of an Engraved Drawing of a Masked Human Figure', *Proceedings of the Prehistoric Society of East Anglia*, 6 (1929), 27–9.
108. Jacobi and Higham, 'The Later Upper Palaeolithic Recolonisation of Britain', p. 230.
109. Currant and Jacobi, 'The Mammal Faunas of the British Late Pleistocene', p. 174.
110. Currant, 'The Lateglacial Mammal Fauna of Gough's Cave', pp. 293–4.
111. Smith, *Late Stone Age Hunters*, p. 98, and R. A. Parkin, P. Rowley-Conwy, and Dale Serjeantson, 'Late Palaeolithic Exploitation of Horse and Red Deer at Gough's Cave, Cheddar, Somerset', *Proceedings of the University of Bristol Spelæological Society*, 17 (1986), 311–30 (pp. 314–16).
112. Jacobi, 'The Late Upper Palaeolithic Lithic Collection from Gough's Cave', p. 77.
113. Tom Lord and John Howard, 'Cave Archaeology', in *Caves and Karst of the Yorkshire Dales*, ed. by Tony Waltham and David Lowe, 2 vols (Buxton: British Cave Research Association, 2013–17), I (2013), pp. 239–51 (p. 241).
114. T. C. Lord et al., 'People and Large Carnivores as Biostratinomic Agents in Lateglacial Cave Assemblages', *Journal of Quaternary Science*, 22 (2007), 681–94 (pp. 689–91).
115. Ibid., p. 689.
116. J. Wilfrid Jackson, 'Report on the Animal Remains Found in the Kilgreany Cave, Co. Waterford', *Proceedings of the University of Bristol Spelæological Society*, 3 (1929), 137–53 (p. 144).
117. Lord et al., 'People and Large Carnivores', p. 692.
118. Montgomery et al., 'Origin of British and Irish Mammals', p. 157.
119. F. E. Mayle et al., 'Climate Variations in Britain During the Last Glacial–Holocene Transition (15.0–11.5 cal ka BP): Comparison with the GRIP Ice-core Record', *Journal of the Geological Society, London*, 156 (1999), 411–23 (p. 415).
120. It is thought that Doggerland became fully submerged somewhere between 8,000 and 7,000 or so years ago when rising sea levels, potentially coupled with a devasting tsunami, flooded the land.
121. A. J. Stuart, 'Insularity and Quaternary Vertebrate Faunas in Britain and Ireland', *Geological Society, London, Special Publications*, 96 (1995), 111–25 (p. 122).
122. L. David Mech and Luigi Boitani, 'Introduction', in *Wolves: Behavior, Ecology, and Conservation*, ed. by Mech and Boitani, pp. xv–xvii (p. xv).
123. Mallory, *Origins of the Irish*, pp. 34–5.
124. Ibid., citing Ceiridwen J. Edwards and Daniel G. Bradley, 'Human Colonisation Routes and the Origins of Irish Mammals', in *Mesolithic Horizons: Papers Presented at the Seventh International Conference on the Mesolithic in Europe, Belfast 2005*, ed. by Sinéad McCartan et al., 2 vols (Oxford: Oxbow, 2009), I, pp. 217–24 (p. 220).
125. Mallory, *Origins of the Irish*, p. 43. See also Edwards and Bradley, 'Human Colonisation Routes and the Origins of Irish Mammals', pp. 221–2.

Chapter 2: Wolves at the Dawn of History

1. George Monbiot, *Feral: Rewilding the Land, Sea and Human Life* (London: Penguin, 2014), pp. 38–9.
2. D. W. Yalden, 'Mammals in Britain – A Historical Perspective', *British Wildlife*, 14.4 (2003), pp. 243–51 (p. 245).
3. R. Coard and A. T. Chamberlain, 'The Nature and Timing of Faunal Change in the British Isles across the Pleistocene/Holocene Transition', *The Holocene*, 9 (1999), 372–6 (pp. 374–5); Yalden, 'Mammals in Britain', p. 245; and Montgomery et al., 'Origin of British and Irish Mammals', p. 157.
4. S. Maroo and D. W. Yalden, 'The Mesolithic Mammal Fauna of Great Britain', *Mammal Review*, 30 (2000), 243–48, and Derek W. Yalden, 'Historical Dichotomies in the Exploitation of Mammals', in *The Exploitation of Mammal Populations*, ed. by Victoria J. Taylor and Nigel Dunstone (London: Chapman and Hall, 1996), pp. 16–27 (p. 22).
5. Graeme Warren, *Hunter-Gatherer Ireland: Making Connections in an Island World* (Oxford: Oxbow Books, 2022), p. 55.
6. Montgomery et al., 'Origin of British and Irish Mammals', p. 149 (Table 1a).
7. Ibid., p. 157, and D. P. Sleeman, 'Quantifying the Prey Gap for Ireland', *The Irish Naturalists' Journal*, 29 (2008), 77–82 (pp. 79–80).
8. Sleeman, 'Quantifying the Prey Gap for Ireland', p. 78.
9. Montgomery et al., 'Origin of British and Irish Mammals', p. 149 (Table 1a).
10. Lord et al., 'People and Large Carnivores', p. 693.
11. Louise H. van Wijngaarden-Bakker, 'The Animal Remains from the Beaker Settlement at Newgrange, Co. Meath: First Report', *Proceedings of the Royal Irish Academy: Archaeology, Culture, History, Literature*, 74 (1974), 313–83 (pp. 340–1).
12. Peter Woodman, *Ireland's First Settlers: Time and the Mesolithic* (Oxford and Philadelphia: Oxbow Books, 2015), p. 27.
13. Carmel McCaffrey and Leo Eaton, *In Search of Ancient Ireland: The Origins of the Irish, From Neolithic Times to the Coming of the English* (Chicago: New Amsterdam Books, 2002), p. 8.
14. Smith, *Late Stone Age Hunters of the British Isles*, pp. 128–9, and Woodman, *Ireland's First Settlers*, p. 267.
15. Hickey, *Wolves in Ireland*, pp. 24–5.
16. Montgomery et al., 'Origin of British and Irish Mammals', p. 157.
17. Warren, *Hunter-Gatherer Ireland*, pp. 57–8.
18. Ibid., pp. 60–1, and Graeme Warren et al., 'The Potential Role of Humans in Structuring the Wooded Landscapes of Mesolithic Ireland: A Review of Data and Discussion of Approaches', *Vegetation History and Archaeobotany*, 23 (2014), 629–46 (p. 632).
19. Miki Ben-Dor and Ran Barkai, 'The Evolution of Paleolithic Hunting Weapons: A Response to Declining Prey Size', *Quaternary*, 6.3 (2023), 1–20 (pp. 13–14).
20. Nick J. Overton and Barry Taylor, 'Humans in the Environment: Plants, Animals and Landscapes in Mesolithic Britain and Ireland', *Journal of World Prehistory*, 31 (2018), 385–402 (p. 388).
21. Ibid.; Naomi Sykes, *Beastly Questions: Animal Answers to Archaeological Issues* (London: Bloomsbury, 2014), p. 57; and Lynne Bevan, 'Stag Nights and Horny Men: Antler Symbolism and Interaction with the Animal World During the Mesolithic', in *Peopling the Mesolithic in a Northern Environment*, ed. by Lynne Bevan and Jenny Moore (Oxford: BAR Publishing, 2016), pp. 35–44 (p. 38).
22. Francis Pryor, *Scenes from Prehistoric Life: From the Ice Age to the Coming of the Romans* (London: Head of Zeus, 2021), p. 34.

23. Nicky Milner, Chantal Conneller, and Barry Taylor, *Star Carr Volume 1: A Persistent Place in a Changing World* (York: White Rose University Press, 2018), pp. 228, 8, and xix.
24. Ibid., p. 270.
25. Derek Yalden, *The History of British Mammals* (London: T & A D Poyser, 1999), p. 65, and Milner, Conneller, and Taylor, *Star Carr Volume 1*, pp. 248–9.
26. Milner, Conneller, and Taylor, *Star Carr Volume 1*, p. 247, and Seamus Caulfield, 'Star Carr: An Alternative View', *Irish Archaeological Research Forum*, 5 (1978), 15–22 (p. 21).
27. Chantal Conneller, 'Becoming Deer: Corporeal Transformations at Star Carr', *Archaeological Dialogues*, 11 (2004), 37–56 (pp. 37 and 42–4).
28. Ibid., pp. 48 and 45.
29. Ibid, p. 48.
30. Milner, Conneller, and Taylor, *Star Carr Volume 1*, p. 343.
31. Ibid., pp. 232 and 256.
32. Ibid., p. 270.
33. C. J. Conneller, 'Star Carr Recontextualised', in *Peopling the Mesolithic in a Northern Environment*, ed. by Bevan and Moore, pp. 81–6 (p. 84).
34. Milner, Conneller, and Taylor, *Star Carr Volume 1*, p. 270.
35. Nick J. Overton, 'More than Skin Deep: Reconsidering Isolated Remains of "Fur-Bearing Species" in the British and European Mesolithic', *Cambridge Archaeological Journal*, 26 (2016), 561–78 (p. 568).
36. Overton and Taylor, 'Humans in the Environment', pp. 396–7.
37. Milner, Conneller, and Taylor, *Star Carr Volume 1*, p. 270.
38. Nicholas J. Overton, 'Memorable Meetings in the Mesolithic: Tracing the Biography of Human-Nonhuman Relationships in the Kennet and Colne Valleys with Social Zooarchaeology' (unpublished doctoral thesis, University of Manchester, 2014), pp. 230–1 and 246.
39. Ibid., pp. 301 and 303.
40. Shipman, *The Invaders*, pp. 194 and 201.
41. This transformation was a protracted process, such that it is difficult to speak of either 'wolves' or 'dogs' when referring to the canids in the early stages of domestication. Estimates of when domestication occurred are highly varied, and while some researchers suggest that the genetic divergence of dogs from wolves could have happened as many as 100,000 years ago (C. Vilà et al., 'Multiple and Ancient Origins of the Domestic Dog', *Science*, 13.276 (1997), 1687–9), the earliest archaeological evidence of a potential 'proto-dog', from Siberia, is around 33,000 years old; Nikolai D. Ovodov et al., 'A 33,000-Year-Old Incipient Dog from the Altai Mountains of Siberia: Evidence of the Earliest Domestication Disrupted by the Last Glacial Maximum', *PLoS One*, 6.7 (2011). Various studies have proposed numerous locations at which domestication first occurred, including Europe, East Asia, the Middle East, Central Asia, and Siberia, while others have found that humans forged a significant relationship with wolves in several different places, at different times, and with different wolf subspecies; Anders Bergström et al., 'Grey Wolf Genomic History Reveals a Dual Ancestry of Dogs', *Nature*, 607 (2022), 313–20 (p. 317). The oldest categorical dog remains (based on 'distinct morphology'), have been found in France, Spain, and Germany, and are between 15,000 and 13,500 years old; Olaf Thalmann and Angela R. Perri, 'Paleogenomic Inferences of Dog Domestication', in *Paleogenomics: Genome-Scale Analysis of Ancient DNA*, ed. by Charlotte Lindqvist and Om P. Rajora (Cham: Springer, 2019), pp. 273–306 (p. 277).
42. James A. Serpell, 'Commensalism or Cross-Species Adoption? A Critical Review of Theories of Wolf Domestication', *Frontiers in Veterinary Science*, 8 (2021), and

Yury E. Herbeck et al., 'Fear, Love, and the Origins of Canid Domestication: An Oxytocin Hypothesis', *Comprehensive Psychoneuroendocrinology*, 9 (2022).

43. Lee Alan Dugatkin, 'The Silver Fox Domestication Experiment', *Evolution: Education and Outreach*, 11 (2018).
44. Shipman, *The Invaders*, p. 197.
45. Ibid., pp. 187–9; Pat Shipman, 'How We Hounded Out the Neanderthals', *New Scientist*, 225.3012 (2015), pp. 26–7 (p. 26); Marie-Pierre Horard-Herbin, Anne Tresset, and Jean-Denis Vigne, 'Domestication and Uses of the Dog in Western Europe from the Paleolithic to the Iron Age', *Animal Frontiers*, 4.1 (2014), 23–31 (p. 27); and Luc Janssens et al., 'A New Look at an Old Dog: Bonn-Oberkassel Reconsidered', Journal of Archaeological Science 92 (2018), 126–38 (p. 127).
46. Francis Pryor, *Britain BC: Life in Britain and Ireland Before the Romans* (London: Harper Perennial, 2004), p. 120.
47. Malcolm Lillie, *Hunters, Fishers and Foragers in Wales: Towards a Social Narrative of Mesolithic Lifeways* (Oxford: Oxbow Books, 2015), p. 264.
48. It is not clear when domestic dogs arrived in Britain and Ireland, nor whether they were brought over from the Continent or domesticated locally. While Star Carr and Mount Sandel represent the oldest sites at which remains of *Canis familiaris* have been found on these islands, it is possible that these animals represent only the evolutionary stage at which these animals become physiologically distinct enough to declare them a new species – early proto-dogs, as we saw in the previous chapter, may already have been present in Britian by the end of the Pleistocene; see pp. 31–2 above. It is certainly possible that 'some sort of "special relationship" was [...] already developing between wolf and man' in Britain and/or Ireland by the end of the Ice Age or beginning of the Holocene; John B. Campbell, *The Upper Palaeolithic of Britain* (Oxford: Clarendon, 1977), p. 129.
49. Warren, *Hunter-Gatherer Ireland*, p. 60.
50. Shipman, *The Invaders*, pp. 189–90.
51. Warren, *Hunter-Gatherer Ireland*, pp. 60–1.
52. Kate M. Clark, 'Dogs and Wolves in the Neolithic of Britain', in *Animals in the Neolithic of Britain and Europe*, ed. by Dale Serjeantson and David Field (Oxford: Oxbow Books, 2006), pp. 33–42 (pp. 34–9).
53. Shipman, *The Invaders*, p. 231.
54. Melinda A. Zeder, 'The Origins of Agriculture in the Near East', *Current Anthropology*, 52.S4 (2011), S221–35 (p. S230), and Nerissa Russell, 'Livestock of the Early Farmers', in *Ancient Europe, 8000 B.C.–A.D. 1000: Encyclopedia of the Barbarian World*, ed. by Peter Bogucki and Pam J. Crabtree, 2 vols (New York, NY: Charles Scibner's Sons, 2004), I: *The Mesolithic to Copper Age (c. 8000–2000 B.C.)*, pp. 211–17 (pp. 213–15). Crop domestication also developed independently in various other centres in South and Central America, Asia, and Africa.
55. Michael Parker Pearson, *Bronze Age Britain*, rev. edn (London: Batsford, 2005), p. 11.
56. Sally Coulthard, *A Short History of the World According to Sheep* (London: Head of Zeus, 2020), p. 6.
57. Pryor, *Britain BC*, p. 120.
58. Parker Pearson, *Bronze Age Britain*, pp. 13–14.
59. Pryor, *Scenes from Prehistoric Life*, pp. 76–7.
60. Overton and Taylor, 'Humans in the Environment', p. 388, and Pryor, *Scenes from Prehistoric Life*, p. 34.
61. O'Connor, 'British Fauna in a Changing World', p. 4.
62. It was not until around 5,000 years ago that the 'wooliness' trait was bred into sheep. Until then they were brown-haired, only growing a woolly undercoat in

the winter that was soon cast off in the spring; Russell, 'Livestock of the Early Farmers', p. 214.

63. Parker Pearson, *Bronze Age Britain*, p. 11.
64. Oliver Rackham, *The History of the Countryside* (London: Weidenfeld & Nicolson, 2020), p. 72.
65. Pryor, *Scenes from Prehistoric Life*, pp. 71–2 and 107.
66. Rackham, *History of the Countryside*, pp. 72–3.
67. Dominic Couzens et al., *Britain's Mammals: A Field Guide to the Mammals of Britain and Ireland* (Princeton, NJ: Princeton University Press, 2017), p. 13.
68. Mark Haughton, Marie Louise Stig Sørensen, and Lise Bender Jørgensen, 'Bronze Age Woollen Textile Production in England: A Consideration of Evidence and Potentials', *Proceedings of the Prehistoric Society*, 87 (2021), 173–88 (p. 175).
69. Malcolm Drew Donalson, *The History of the Wolf in Western Civilization: From Antiquity to the Middle Ages* (Lewiston, NY: Mellen, 2006), p. 2.
70. Mary Midgley, 'The Problem of Living with Wildness', in *Wolves and Human Communities: Biology, Politics and Ethics*, ed. by Virginia A. Sharpe, Bryan Norton, and Strachan Donnelley (Washington, DC: Island Press, 2001), pp. 179–90 (p. 180).
71. Mel Davies, 'Cave Archaeology in North Wales', in *Limestones and Caves of Wales*, ed. by Trevor D. Ford (Cambridge: Cambridge University Press, 1989), pp. 92–101 (pp. 98–9).
72. Fiona Beglane, 'Prehistoric Perforated and Worked Animal Teeth', in *Underground Archaeology: Studies on Human Bones and Artefacts from Ireland's Caves*, ed. by Marion Dowd (Oxford: Oxbow Books, 2016), pp. 120–7 (pp. 121–2).
73. Ibid., p. 124.
74. Marion Dowd, *The Archaeology of Caves in Ireland* (Oxford: Oxbow Books, 2015), pp. 119–20.
75. Lord and Howard, 'Cave Archaeology', pp. 244–5.
76. Joshua Pollard, 'A Community of Beings: Animals and People in the Neolithic of Southern Britain', in *Animals in the Neolithic of Britain and Europe*, ed. by Serjeantson and Field, pp. 135–48 (p. 140).
77. J. N. Graham Ritchie, 'The Stones of Stenness, Orkney', *Proceedings of the Society of Antiquaries of Scotland*, 107 (1976), 1–60 (p. 10), and G. J. Barclay, 'The "Henge" and "Hengiform" in Scotland', in *Set in Stone: New Approaches to Neolithic Monuments in Scotland*, ed. by Vicki Cummings and Amelia Pannett (Oxford: Oxbow, 2005), pp. 81–94 (p. 91).
78. Benjamin W. Roberts, 'Britain and Ireland in the Bronze Age: Farmers in the Landscape or Heroes on the High Seas?', in *The Oxford Handbook of the European Bronze Age*, ed. by Harry Fokkens and Anthony Harding (Oxford: Oxford University Press, 2013), pp. 531–49 (p. 538).
79. Beglane, 'Prehistoric Perforated and Worked Animal Teeth', pp. 121–4.
80. Marion Dowd, 'Artefacts and Bones from Glencurran Cave', *Burren Insight* (2010), 10–12 (p. 10).
81. John Waddell, *The Prehistoric Archaeology of Ireland* (Galway: Galway University Press, 1998), pp. 148 and 165 n. 64, and Dowd, *Archaeology of Caves*, pp. 154–6.
82. Dowd, *Archaeology of Caves*, p. 119.
83. Ibid., p. 113, and Beglane, 'Prehistoric Perforated and Worked Animal Teeth', p. 125.
84. Martin Smith, 'Bones Chewed by Canids as Evidence for Human Excarnation: A British Case Study, *Antiquity*, 80 (2006), pp. 671–85 (pp. 681–2).

85. John Denton Blore, 'Lynx Cave, Denbighshire: 50 Years of Excavation, 1962–2012' (Wallasey: printed by the author, 2012), pp. 35, 42, and 50.
86. Smith, 'Bones Chewed by Canids', p. 671.
87. Ibid., p. 683.
88. Joshua Pollard, 'A Community of Beings: Animals and People in the Neolithic of Southern Britain', in *Animals in the Neolithic of Britain and Europe*, ed. by Serjeantson and Field, pp. 135–48 (p. 140).
89. Ibid., pp. 139–40.
90. Dowd, *Archaeology of Caves*, p. 113.
91. It is not known precisely when the Celts arrived in Britain and Ireland, nor from where the culture and people stemmed. While it was previously suggested that the introduction of Celtic culture to Britain and Ireland was due to mass invasion of Celtic peoples from the European continent at some point during the Iron Age, a 2022 study found that migration from Europe to Britain over several centuries was responsible for the introduction of Celtic culture and language at the end of the Bronze Age. Conversely, this study found little evidence of mass migration in the Iron Age; Nick Patterson et al. 'Large-scale Migration into Britain During the Middle to Late Bronze Age', *Nature*, 601 (2022), 588–94. The nature of the arrival of Celtic culture and language in Ireland is a complex issue, but is similarly indicative of integration rather than mass invasion; Katharina Becker, 'Iron Age Ireland: Continuity, Change, and Identity', in *Atlantic Europe in the First Millennium* BC*: Crossing the Divide*, ed. by Tom Moore and Xosé-Lois Armada (Oxford: Oxford University Press, 2011), pp. 449–67.
92. Miranda Green, *Animals in Celtic Life and Myth* (London and New York: Routledge, 2002), pp. 1–4.
93. Colin Haselgrove, 'The Iron Age', in *The Archaeology of Britain*, ed. by Hunter and Ralston, pp. 149–74 (pp. 149 and 168–9).
94. Green, *Animals in Celtic Life and Myth*, p. 5.
95. Ibid., pp. 7–8 and 11; Hickey, *Wolves in Ireland*, p. 26; and James Fairley, *An Irish Beast Book: A Natural History of Ireland's Furred Wildlife* (Belfast: Blackstaff, 1984), pp. 291–2.
96. Dara Sands, 'Dewilding "Wolf-land"', *Conservation & Society*, 20 (2022), 257–67 (p. 260).
97. Robert Winder, *The Last Wolf: The Hidden Springs of Englishness* (London: Little, Brown, 2017), p. 32.
98. Finbar McCormick and Paul C. Buckland, 'The Vertebrate Fauna', in *Scotland After the Ice Age: Environment, Archaeology and History 8000* BC *–* AD *1000*, ed. by Kevin J. Edwards and Ian B. M. Ralston (Edinburgh: Edinburgh University Press, 1997), pp. 83–103 (p. 100), and Yalden, *History of British Mammals*, pp. 229–30.
99. Miranda Aldhouse-Green, *An Archaeology of Images: Iconology and Cosmology in Iron Age and Roman Europe* (London and New York: Routledge, 2005), p. 126.
100. O'Connor, 'British Fauna in a Changing World', p. 6.
101. See Shimon Applebaum, 'Agriculture in Roman Britain', *The Agricultural History Review*, 6 (1958), 66–86 (pp. 74–81).
102. Finbar McCormick, 'Agriculture, Settlement and Society in Early Medieval Ireland', *Quaternary International*, 346 (2014), 119–30 (p. 119).
103. Joan P. Alcock, *A Brief History of Roman Britain* (London: Constable & Robinson, 2011), p. 312.
104. Mauro Rizzetto, Pam J. Crabtree, and Umberto Albarella, 'Livestock Changes at the Beginning and End of the Roman Period in Britain: Issues of Acculturation,

Adaptation, and "Improvement"', *European Journal of Archaeology*, 20 (2017), 535–56 (p. 550).

105. Rackham, *History of the Countryside*, p. 74.
106. Ibid., p. 383.
107. Ibid., pp. 122–3.
108. Hector Boece, *Scotorum Historia (1575 version): A Hypertext Critical Edition*, ed. and trans. by Dana F. Sutton (Birmingham: The Philological Museum, 2010), Books 3 and 2 respectively <https://philological.cal.bham.ac.uk/boece> [accessed 13 January 2025].
109. Sam Milton Wilford, 'Riddles in the Dark? The Human Use of Caves During the 1st Millennia BC and AD Across the British Isles' (unpublished doctoral thesis, Durham University, 2016), p. 475.
110. Green, *Animals in Celtic Life and Myth*, pp. 54–5.
111. Anna Gannon, *The Iconography of Early Anglo-Saxon Coinage: Sixth to Eighth Centuries* (Oxford: Oxford University Press, 2003), p. 129.
112. Miranda Aldhouse-Green, *Boudica Britannia* (London and New York: Routledge, 2014), pp. 23–4.
113. John T. Koch, 'Cymru (Wales)', in *Celtic Culture: An Historical Encyclopedia*, ed. by John T. Koch, 5 vols (Santa Barbara, CA: ABC-CLIO, 2006), II, pp. 529–32 (p. 532), and John T. Koch, *Celto-Germanic, Later Prehistory and Post-Proto-Indo-European Vocabulary in the North and West* (Aberystwyth: University of Wales Centre for Advanced Welsh and Celtic Studies, 2020), p. 96.
114. Daphne Nash Briggs, 'Reading the Images on Iron Age Coins: 3. Some Cosmic Wolves', *Chris Rudd List*, 110 (2010), pp. 2–4.
115. Martin Henig, *Religion in Roman Britain* (London: Batsford, 2005), p. 49.
116. Joan P. Alcock, 'Three Bronze Figurines in the British Museum', *Antiquaries Journal*, 43 (1963), 118–23 (p. 122).
117. Emma Durham, 'Style and Substance: Some Metal Figurines from South-West Britain', *Britannia*, 45 (2014), 195–221 (pp. 207–10 and p. 198 (Figure 3)).
118. Ibid., pp. 209–10.
119. J. A. Macculloch, *The Religion of the Ancient Celts* (London: Constable, 1991), p. 218.
120. Tom Sjöblom, 'The Great Mother: The Cult of the Bear in Celtic Traditions', *Studia Celtica Fennica*, 3 (2006), 71–8 (p. 74), and Robin Melrose, *Religion in Britain from the Megaliths to Arthur: An Archaeological and Mythological Exploration* (Jefferson, NC: McFarland & Co., 2016), p. 165.
121. Ralph Haussler, '*Apollo Cunomaglos*, Lord of the Wolves', *Bandue*, 11 (2018–19), 65–82.
122. Martin Henig, *The Art of Roman Britain* (London: Batsford, 2002), p. 55.
123. Ashmolean Museum, 'British Collections by Archaeological Period: Roman (AD 43 - AD 410)' <https://britisharchaeology.ashmus.ox.ac.uk/collections/roman.html> [accessed 20 October 2025], and Mika Rissanen, 'The *Lupa Romana* in the Roman Provinces', *Acta Archaeologica Academiae Scientiarum Hungaricae* 65 (2014), pp. 335–60 (p. 343 (Figure 7)).
124. Catherine Johns, *Dogs: History, Myth, Art* (London: British Museum Press, 2008), p. 105.
125. Mika Rissanen, 'Was There a Taboo on Killing Wolves in Rome?', *Quaderni Urbinati di Cultura Classica*, 107 (2014), 125–47 (pp. 143–4, 141, 126, 130, and 127).
126. Ibid., pp. 131–2.
127. Ibid., p. 135.
128. Jonathan Faiers, *Fur: A Sensitive History* (New Haven, CT: Yale University Press, 2020), p. 82.

129. Rissanen, 'Was There a Taboo on Killing Wolves in Rome?', pp. 139 and 144.
130. Laura D. Gelfand, 'The Wolf at the Door and the Dog at Our Feet', *Home Cultures*, 18 (2021), 105–27 (p. 112).
131. Rissanen, 'Was There a Taboo on Killing Wolves in Rome?', p. 136.
132. Luigi Boitani, 'Ecological and Cultural Diversities in the Evolution of Wolf-human Relationships', in *Ecology and Conservation of Wolves in a Changing World*, ed. by Ludwig N. Carbyn, Steven H. Fritts, and Dale R. Seip (Edmonton: Canadian Circumpolar Institute, 1995), pp. 3–12 (p. 7).

Chapter 3: Woden and the Wolf-warriors

1. Dál Riata and the Pictish kingdoms merged to form the Kingdom of Alba in the ninth century.
2. Because of its politicisation and appropriation by far-right groups, who uphold an 'imagined Anglo-Saxon heritage as an exemplar of European whiteness' (Mary Rambaran-Olm and Erik Wade, 'The Many Myths of the Term 'Anglo-Saxon', *Smithsonian*, 14 July 2021 <https://www.smithsonianmag.com/history/many-myths-term-anglo-saxon-180978169>), the term 'Anglo-Saxon' (which was not actually used by the people themselves) is now generally being avoided by scholars of the period.
3. Mateusz Fafinski, *Roman Infrastructure in Early Medieval Britain: The Adaptations of the Past in Text and Stone* (Amsterdam: Amsterdam University Press, 2021), p. 93.
4. Pryor, *Scenes from Prehistoric Life*, pp. 291–2, and Fafinski, *Roman Infrastructure in Early Medieval Britain*, pp. 90–1 and 84.
5. Fafinski, *Roman Infrastructure in Early Medieval Britain*, p. 101.
6. Hannah J. O'Regan, 'The Presence of the Brown Bear *Ursus arctos* in Holocene Britain: A Review of the Evidence', *Mammal Review*, 48 (2018), 229–44.
7. Nigel T. Monaghan, 'The Brown Bear (*Ursus arctos* L.) in Ireland', *Irish Naturalists' Journal*, 40 (2023), 1–19 (p. 15).
8. David A. Hetherington, Tom C. Lord, and Roger M. Jacobi, 'New Evidence for the Occurrence of Eurasian Lynx (*Lynx lynx*) in Medieval Britain', *Journal of Quaternary Science*, 21 (2006), 3–8.
9. Lee Raye, *The Atlas of Early Modern Wildlife: Britain and Ireland Between the Middle Ages and the Industrial Revolution* (London: Pelagic Publishing, 2023), pp. 39–40.
10. Aleksander Pluskowski, *Wolves and the Wilderness in the Middle Ages* (Woodbridge: Boydell, 2006), p. 117.
11. See p. 84 below.
12. Sam Leggett and Tom Lambert, 'Food and Power in Early Medieval England: A Lack of (Isotopic) Enrichment', *Anglo-Saxon England*, 48 (2022), 1–42.
13. Pluskowski, *Wolves and the Wilderness*, p. 115.
14. Abigail Firey, *A Contrite Heart: Prosecution and Redemption in the Carolingian Empire* (Leiden: Brill, 2009), p. 236, and *Canones Adomnani* ('The Canons of Adamnan', in *The Irish Penitentials*, ed. by Ludwig Bieler (Dublin: Dublin Institute for Advanced Studies, 1963), pp. 176–81 (p. 181).
15. Audrey L. Meaney, *Anglo-Saxon Amulets and Curing Stones* (Oxford: B. A. R., 1981), p. 138.
16. Pluskowski, *Wolves and the Wilderness*, p. 115.
17. *Leechdoms, Wortcunning and Starcraft of Early England*, ed. by Oswald Cockayne, 3 vols (London: Longman, Green, Longman, Roberts, and Green, 1864–6), I (1864), pp. 361–3. It is possible that 'wolf's milk' refers not the milk

from a wolf, however, but to a type of spurge plant with milky-looking juice; Hana Videen, *The Deorhord: An Old English Bestiary* (Princeton, NJ: Princeton University Press, 2024), p. 211.

18. See *Leechdoms, Wortcunning and Starcraft*, ed. by Cockayne, II (1865), pp. 103 and 133.
19. See *Leechdoms, Wortcunning and Starcraft*, ed. by Cockayne, I (1864), p. 361.
20. Donald K. Fry, '*Wulf and Eadwacer*: A Wen Charm', *Chaucer Review*, 5 (1971), 247–63 (p. 251), and Meaney, *Anglo-Saxon Amulets and Curing Stones*, pp. 19–20.
21. James Morris, 'Red Deer's Role in Social Expression on the Isles of Scotland', in *Just Skin and Bones? New Perspectives on Human-animal Relations in the Historical Past*, ed. by Aleksander Pluskowski (Oxford: Archaeopress, 2005), pp. 9–18; Pluskowski, *Wolves and the Wilderness*, p. 59; and Fiona Beglane, 'Deer in Medieval Ireland: Preliminary Evidence from Kilteasheen, Co. Roscommon', in *Medieval Lough Ce: History, Archaeology and Landscape*, ed. by Thomas Finan (Dublin: Four Courts Press, 2010), pp. 145–58.
22. Pluskowski, *Wolves and the Wilderness*, p. 59, and Beglane, 'Deer in Medieval Ireland: Preliminary Evidence from Kilteasheen', p. 151.
23. Pluskowski, *Wolves and the Wilderness*, pp. 59–60, and Della Hooke, 'Pre-Conquest Woodland: Its Distribution and Usage', *The Agricultural History Review*, 37 (1989), 113–29 (p. 125).
24. Richard Jones, *The Medieval Natural World* (London: Routledge, 2013), p. 78.
25. Elizabeth Marshall, *Wolves in Beowulf and Other Old English Texts* (Woodbridge: Brewer, 2022), p. 4.
26. Yalden, *History of British Mammals*, p. 163.
27. Debby Banham and Rosamond Faith, *Anglo-Saxon Farms and Farming* (Oxford: Oxford University Press, 2014), p. 118, and Flight, *Basilisks and Beowulf*, pp. 70–1.
28. Pluskowski, *Wolves and the Wilderness*, p. 79.
29. Trevor Rowley, *Landscapes of the Norman Conquest* (Barnsley: Pen & Sword Archaeology, 2022), p. 16.
30. Videen, *The Deorhord*, p. 46.
31. Nancy Edwards, *Life in Early Medieval Wales* (Oxford: Oxford University Press, 2023), pp. 205–7.
32. McCormick, 'Agriculture, Settlement and Society in Early Medieval Ireland', pp. 119 and 120. See pp. 51–2 above.
33. Mac Coitir, *Ireland's Animals*, p. 12, and Clare Downham, *Medieval Ireland* (Cambridge: Cambridge University Press, 2018), p. 31.
34. Downham, *Medieval Ireland*, pp. 31–2.
35. Mac Coitir, *Ireland's Animals*, pp. 50–1.
36. Edouard Masson-MacLean et al., 'New Zooarchaeological Evidence from Pictish Sites in Scotland: Implications for Early Medieval Economies and Animal–Human Relationships', *Frontiers in Environmental Archaeology*, 2 (2023), 1–10 (p. 7).
37. Aleksander Pluskowski, 'Wolves and Sheep in Medieval Semiotics, Iconology and Ecology: A Case Study of Multi- and Inter-disciplinary Approaches to Human–Animal Relations in the Historical Past', in *Animal Diversities*, ed. by Gerhard Jaritz and Alice Choyke (Krems: Medium Aevum Quotidianum, 2005), pp. 9–22 (p. 15).
38. Flight, *Basilisks and Beowulf*, pp. 69–70.
39. Marshall, *Wolves in Beowulf*, p. 5, translating *Ælfric's 'Colloquy'*, ed. by G. N. Garmonsway, rev. edn (Exeter: University of Exeter Press, 1978), p. 22.
40. Aybes and Yalden, 'Place-name Evidence', p. 205.

41. Ian B. M. Ralston and Ian Armit, 'The Early Historic Period: An Archaeological Perspective', in *Scotland After the Ice Age*, ed. by Edwards and Ralston, pp. 217–40 (p. 232).
42. *The Metrical Dindshenchas*, ed. and trans. by Edward Gwynn, Corpus of Electronic Texts Edition, comp. by Lisa Boucher, Alf Siewers, and Saorla Ó Corráin <https://celt.ucc.ie/published/T106500D> [accessed 08 December 2024].
43. Fergus Kelly, *Early Irish Farming: A Study Based Mainly on the Law-texts of the 7th and 8th Centuries* (Dublin: Dublin Institute for Advanced Studies, 2000), p. 187.
44. Niall Mac Coitir, *Ireland's Animals: Myths, Legends and Folklore* (Cork: Collins Press, 2015), p. 50.
45. *Ancient Laws of Ireland*, ed. by W. Neilson Hancock and others, 6 vols (Dublin: Thom, 1865–1901), I: *Introduction to Senchus Mor, and Athgabail; or, Law of Distress* (1865), p. 161.
46. *Ancient Laws of Ireland*, ed. by Hancock and others, II: *Senchus Mor, Part II*, ed. by W. Neilson Hancock and Thaddeus O'Mahony (1869), p. 271.
47. William of Malmesbury, *Gesta Regum Anglorum: History of the English Kings*, ed. and trans. by R. A. B. Mynors, R. M. Thomson, and M. Winterbottom (Oxford: Clarendon, 1998), I, p. 255.
48. *Holinshed's Chronicles of England, Scotland, and Ireland*, 6 vols (London: Johnson, Rivington, Payne, Wilkie and Robinson, Longman, Hurst, Rees and Orme, Cadell and Davies, and Mawman, 1807–8), I: England (1807), p. 378.
49. C. Bohun Smyth, 'The First and Last Days of the Saxon Rule in Sussex', *Sussex Archaeological Collections*, 4 (1851), 67–92 (p. 83), cited in Harting, *British Animals*, p. 132.
50. Derek Gow, *Hunt for the Shadow Wolf: The Lost History of Wolves in Britain and the Myths and Stories that Surround Them* (London: Chelsea Green, 2024), p. 39.
51. Cledwyn Fychan, *Galwad y Blaidd: Perthynas y Blaidd â Chymru dros y Canrifoedd* (Aberystwyth: Ceredigion, 2006), p. 48.
52. Tim Flight, 'The Wolf Must Be in the Woods: The Real and Mythical Dangers of the Wilderness', *History Today*, 67.6 (2017), 18–20 (p. 20).
53. Aleksander Pluskowski, 'Where Are the Wolves? Investigating the Scarcity of European Grey Wolf (*Canis lupus lupus*) Remains in Medieval Archaeological Contexts and Its Implications', *International Journal of Osteoarchaeology*, 16 (2006), 279–95 (p. 289), and Yalden, *History of British Mammals*, p. 132.
54. Robert E. Stillman, 'Philip Sidney, Thomas More, and Table Talk: Texts/Contexts', *English Literary Renaissance*, 45 (2015), 323–50 (p. 346).
55. Malmesbury, *Gesta Regum Anglorum*, ed. and trans. by Mynors, Thomson, and Winterbottom, I, p. 255.
56. Pluskowski, *Wolves and the Wilderness*, p. 98.
57. Ibid., p. 100, and Aybes and Yalden, 'Place-name Evidence', pp. 204–5 and 213 (Table 4).
58. Pluskowski, *Wolves and the Wilderness*, p. 102.
59. Tim Flight, *Basilisks and Beowulf: Monsters in the Anglo-Saxon World* (London: Reaktion, 2021), p. 79.
60. Rackham, *History of the Countryside*, p. 353.
61. See pp. 75–8 below.
62. Pluskowski, *Wolves and the Wilderness*, p. 194.
63. Ibid., p. 80.
64. Ibid., pp. 11 and 46.
65. Flight, *Basilisks and Beowulf*, p. 66.
66. Hooke, 'Pre-Conquest Woodland', p. 122.

67. Pluskowski, *Wolves and the Wilderness*, pp. 43, 46, and 47, and Hooke, 'Pre-Conquest Woodland', pp. 120 and 113.
68. Pluskowski, *Wolves and the Wilderness*, pp. 47–50.
69. Downham, *Medieval Ireland*, pp. 35, 36, and 41.
70. Ibid., p. 35.
71. Sarah Harlan-Haughey, *The Ecology of the English Outlaw in Medieval Literature: From Fen to Greenwood* (London: Routledge, 2016), p. 26. Translation my own.
72. Marshall, *Wolves in Beowulf*, p. 30.
73. Hooke, 'Pre-Conquest Woodland', p. 113, and Pluskowski, *Wolves and the Wilderness*, p. 44.
74. Pluskowski, *Wolves and the Wilderness*, pp. 49–50 and 47.
75. Aidan O'Sullivan, 'Woodlands', in *Medieval Ireland: An Encyclopedia*, ed. by Seán Duffy (Abingdon: Routledge, 2016), pp. 522–3 (p. 523), and McCormick, 'Agriculture, Settlement and Society in Early Medieval Ireland', p. 119.
76. Pluskowski, *Wolves and the Wilderness*, pp. 32 and 38.
77. Ibid., pp. 35, 44, and 46.
78. Translation my own.
79. Heide Estes, *Anglo-Saxon Literary Landscapes: Ecotheory and the Environmental Imagination* (Amsterdam: Amsterdam University Press, 2017), p. 16.
80. Pluskowski, *Wolves and the Wilderness*, p. 46.
81. Anthony Dent, *Lost Beasts of Britain* (London: Harrap, 1974), p. 128.
82. Ibid., p. 124.
83. Marshall, *Wolves in Beowulf*, pp. 105–6.
84. See ibid., pp. 102–7.
85. Pluskowski, *Wolves and the Wilderness*, p. 194.
86. Ibid., p. 39.
87. Aybes and Yalden, 'Place-name Evidence', p. 204.
88. Ibid., pp. 204, 206–10 (Table 2), 211 (Table 3), and 202–3.
89. Yalden, *History of British Mammals*, pp. 135–6.
90. See p. 109 below.
91. Aybes and Yalden, 'Place-name Evidence', pp. 204–5.
92. Ibid., pp. 204 and 222.
93. Ibid., p. 222.
94. Yalden, *History of British Mammals*, p. 136.
95. Although Cledwyn Fychan lists around 200 Welsh place-names with these elements in his book *Galwad y Blaidd*, many of them may be related to places named for people named Blaidd or Bleddyn, while *cnud* is a term used for both wolf packs and groups of people; *Galwad y Blaidd*, p. 190, and *A Dictionary of the Welsh Language* <https://www.geiriadur.ac.uk/gpc/gpc.html> [accessed 05 January 2025].
96. Stephen Aldhouse-Green, 'The Pontnewydd People, Their Cave and Their World', in *Neanderthals in Wales*, ed. by Aldhouse-Green, Peterson, and Walker, pp. 327–44 (p. 340).
97. Aybes and Yalden, 'Place-name Evidence', p. 214.
98. Hickey, *Wolves in Ireland*, p. 29.
99. Kim McCone, 'Varia II', Ériu, 36 (1985), 171–6 (p. 171), and Kristen Mills, 'Glossing the Glosses: The Right Marginal Notes on *Glaidomuin* and *Gudomhuin* in TCD MS 1337', *Studia Celtica Fennica*, 15 (2018), 65–82 (p. 66).
100. Hickey, *Wolves in Ireland*, pp. 29–30.
101. Ibid., pp. 30 and 32.
102. Ibid., p. 30.
103. Ibid., pp. 30–2.

104. See pp. 51–2 and 68 above.
105. Little is known of the Pictish language, which was replaced by or incorporated into Gaelic as the latter spread across Scotland.
106. Yalden, *History of British Mammals*, p. 136, and Aybes and Yalden, 'Place-name Evidence', p. 211.
107. Andrew E. M. Wiseman, 'A Noxious Pack': Historical, Literary and Folklore Traditions of the Wolf (*Canis Lupus*) in the Scottish Highlands', *Scottish Gaelic Studies*, 25 (2009), 95–142 (p. 99).
108. Aybes and Yalden, 'Place-name Evidence', p. 211.
109. Elizabeth Sutherland, *In Search of the Picts* (London: Constable, 1994), pp. 9–85.
110. Ross Barnett, *The Missing Lynx: The Past and Future of Britain's Lost Mammals* (London: Bloomsbury Wildlife, 2019), p. 258.
111. Ibid.
112. Richard Verstegan, *A Restitution of Decayed Intelligence in Antiquities Concerning the Most Noble and Renowned English Nation* (London: Norton, 1634), p. 59. Spelling modernised.
113. Wiseman, 'A Noxious Pack', p. 99, and Antone Minard, 'Calendar, Celtic', in *Celtic Culture*, ed. by Koch, I, pp. 330–2 (p. 331).
114. Pluskowski, *Wolves and the Wilderness*, p. 102, and Wiseman, 'A Noxious Pack', p. 99.
115. See p. 81 below.
116. Pluskowski, *Wolves and the Wilderness*, p. 102.
117. *Calendar of the Patent Rolls Preserved in the Public Record Office. Henry VI: Volume 5: 1446–1452* (London: Brothers, 1909), p. 69.
118. Flight, *Basilisks and Beowulf*, p. 80.
119. Aleks Pluskowski, 'The Tyranny of the Gingerbread House: Contextualising the Fear of Wolves in Medieval Northern Europe through Material Culture, Ecology and Folklore, *Current Swedish Archaeology*, 13 (2005), 141–60 (p. 151).
120. Jim Crumley, *The Last Wolf* (Edinburgh: Birlinn, 2010), p. 32.
121. s.v. 'loop n.[2]', *Oxford English Dictionary Online* (2024) <https://www.oed.com/dictionary/loop_n2?tab=etymology#38836062>, and s.v. 'loophole n.[1]', *Oxford English Dictionary Online* (2024) <https://www.oed.com/dictionary/loophole_n1?tab=meaning_and_use#38837621> [accessed 08 January 2025].
122. See Elizabeth Marshall, 'A "Wasteland" Infested by Wolves: The Fallacy of "Dark Age" England', in *The Wolf: Culture, Nature, Heritage*, ed. by Ian Convery et al. (Woodbridge: Boydell, 2023), pp. 107–15.
123. Jennifer Neville, *Representations of the Natural World in Old English Poetry* (Cambridge: Cambridge University Press, 1999), p. 6.
124. Pluskowski, *Wolves and the Wilderness*, p. 108.
125. Ibid.
126. John D. C. Linnell et al., 'The Fear of Wolves: A Review of Wolf Attacks on Humans', Norsk institutt for naturforskning Oppdragsmelding, 731 (2002), 1–65 (pp. 4 and 36).
127. Flight, *Basilisks and Beowulf*, pp. 76–7.
128. Marshall, *Wolves in Beowulf*, p. 25.
129. Translation my own.
130. *Aneirin: Y Gododdin, Britain's Oldest Heroic Poem*, ed. and trans. by A. O. H. Jarman (Llandysul: Gomer, 1988), p. 6 (line 62) and p. 22 (line 330).
131. *Homily IV*, trans. by Francis M. Clough, in *The Vercelli Book Homilies: Translations from the Anglo-Saxon*, ed. by Lewis E. Nicholson (Lanham, NY: University Press of America, 1991), pp. 37–45 (p. 44).
132. Wolves are facultative scavengers who are unlikely to pass up an opportunity for free food if it saves them from expending precious energy upon hunting.

They are also intelligent enough to identify armed men and perhaps to associate large groups of them with a glut of free food. This may explain why, in many examples of the Beast of Battle topos, wolves trail armies on their way to battle. It is also common to see modern wolves with one of the other Beasts of Battle, the raven, with whom they share an interesting relationship. Ravens frequently follow wolves so that they can scavenge from their kills, and it is thought that ravens may even lead wolves to injured prey by harassing the unfortunate animal. Perhaps the two species also associated with one another in early medieval Britain and Scandinavia, hence why they are so often depicted together in the Beast of Battle topos; Thomas J. T. Williams, '"For the Sake of Bravado in the Wilderness": Confronting the Bestial in Anglo-Saxon Warfare', in *Representing Beasts in Early Medieval England and Scandinavia*, ed. by Michael D. J. Bintley and Thomas J. T. Williams, Anglo-Saxon Studies, 29 (Woodbridge: Boydell, 2015), pp. 176–204 (pp. 179–80), and Mohamed Eric Rahman Lacey, 'Birds and Bird-lore in the Literature of Anglo-Saxon England' (unpublished doctoral thesis, University College London, 2013), pp. 114–17 and 118–19.

133. Crumley, *The Last Wolf*, p. 62.
134. Pluskowski, *Wolves and the Wilderness*, pp. 108–9, and Linnell et al., 'The Fear of Wolves', p. 37.
135. Flight, *Basilisks and Beowulf*, p. 76.
136. Joyce E. Salisbury, *The Beast Within: Animals in the Middle Ages*, second edn (London: Routledge, 2011), p. 69.
137. 'Pagan' is a catchall word which simply means 'not Christian'. It is not indicative of defined or unified theological concepts, nor organised religion, but a broad range of worldviews and localised beliefs. Animals may have been central to the spiritual beliefs of the pagan Germanic invaders, functioning as 'mediators between the natural and supernatural worlds' and perhaps revered as deities or representations of them; Aleksander Pluskowski, 'Animal Magic', in *Signals of Belief in Early England: Anglo-Saxon Paganism Revisited*, ed. by Martin Carver, Alex Sanmark, and Sarah Semple (Oxford: Oxbow Books, 2010), pp. 103–27 (pp. 115–16).
138. Gildas, *The Ruin of Britain and Other Works*, ed. and trans. by Michael Winterbottom (London and Chichester: Phillimore, 1978), p. 26.
139. Judith Jesch, 'Eagles, Ravens and Wolves: Beasts of Battle, Symbols of Victory and Death', in *The Scandinavians from the Vendel Period to the Tenth Century: An Ethnographic Perspective*, ed. by Judith Jesch (Woodbridge: Boydell, 2002), pp. 251–80 (p. 263), and Pluskowski, *Wolves and the Wilderness*, p. 155.
140. Hilda Ellis Davidson, *The Lost Beliefs of Northern Europe* (London: Routledge, 1993), p. 76.
141. Williams, 'Confronting the Bestial in Anglo-Saxon Warfare', pp. 192–3.
142. Michael P. Speidel, *Ancient Germanic Warriors: Warrior Styles from Trajan's Column to Icelandic Sagas* (New York: Routledge, 2004), p. 26.
143. Rebecca Pinner, *The Cult of St Edmund in Medieval East Anglia* (Woodbridge: Boydell, 2015), p. 222.
144. Sam Newton, *The Origins of Beowulf and the Pre-Viking Kingdom of East Anglia* (Cambridge: Brewer, 1993), p. 106.
145. Stephen O. Glosecki, 'Wolf of the Bees: Germanic Shamanism and the Bear Hero', *Journal of Ritual Studies*, 2 (1988), 31–53 (pp. 43–6), and Stephen O. Glosecki, 'Wolf [Canis lupus] and Werewolf', in *Medieval Folklore, An Encyclopaedia of Myths, Legends, Tales, Beliefs and Customs*, ed. by Carl Lindahl, John McNamara, and John Lindow (Oxford: Oxford University Press, 2002), pp. 440–2 (p. 441).
146. Marijane Osborn, 'Archaic Magic of Wolf and Eagle in the Anglo-Saxon "Wen Charm"', in *The Book of Nature and Humanity in the Middle Ages and*

the Renaissance, ed. by David Hawkes and Richard G. Newhauser (Turnhout: Brepols, 2013), pp. 223–38 (pp. 234–5).
147. Speidel, *Ancient Germanic Warriors*, p. 24.
148. Newton, *Origins of Beowulf*, p. 108.
149. Pluskowski, *Wolves and the Wilderness*, p. 143.
150. Newton, *The Origins of Beowulf*, p. 109.
151. Pluskowski, *Wolves and the Wilderness*, p. 146.
152. Catherine E. Karkov, *The Art of Anglo-Saxon England* (Woodbridge: Boydell, 2011), pp. 106–7.
153. Carol Neuman de Vegvar, 'The Travelling Twins: Romulus and Remus in Anglo-Saxon England', in *Northumbria's Golden Age*, ed. by Jane Hawkes and Susan Mills (Stroud: Sutton, 1999), pp. 256–67 (p. 259).
154. Pinner, *The Cult of St Edmund*, p. 222.
155. Newton, *The Origins of Beowulf*, p. 109.
156. Pinner, *The Cult of St Edmund*, pp. 222–3.
157. Marshall, *Wolves in Beowulf*, p. 26.
158. St Patrick, '*Letter to the Soldiers of Coroticus*', trans. by Padraig McCarthy (Dublin: Royal Irish Academy, 2011) <https://www.confessio.ie/etexts/epistola_english#undefined> [accessed 10 December 2024].
159. T. W. Rolleston, *The High Deeds of Finn and Other Bardic Romances of Ancient Ireland* (London: Harrap, 1910), pp. 174–5.
160. *Marwnad Owein* ('The Elegy of Owein'), in *Poems from the Book of Taliesin*, ed. and trans. by J. Gwenogvryn Evans (Llanbedrog: the author, 1915), pp. 124–5 (lines 17–18).
161. *Y Gododdin*, ed. and trans. by Jarman, p. 4 (line 49) and p. 40 (line 603), and *The 'Gododdin' of Aneirin: Text and Context from Dark-Age North Britain*, ed. by John Thomas Koch (Cardiff: University of Wales Press, 1997), p. 25 (line 1,250).
162. *Y Gododdin*, trans. by Jarman, pp. 4–6 (lines 56–64).
163. Benjamin Hudson, *Macbeth Before Shakespeare* (Oxford: Oxford University Press, 2022), p. 18.
164. Wiseman, 'A Noxious Pack', pp. 99–100.
165. Translations my own.
166. Rachel Bromwich, 'First Transmission to England and France', in *The Arthur of the Welsh: The Arthurian Legend in Medieval Welsh Literature*, ed. by Rachel Bromwich, A. O. H. Jarman, and Brynley F. Roberts (Cardiff: University of Wales Press, 2020), pp. 273–98 (p. 287).
167. Hickey, *Wolves in Ireland*, p. 34.
168. Aleksander Pluskowski, 'Animal Magic', in *Signals of Belief in Early England: Anglo-Saxon Paganism Revisited*, ed. by Martin Carver, Alex Sanmark, and Sarah Semple (Oxford: Oxbow Books, 2010), pp. 103–27 (p. 117).
169. Translations my own.
170. Peter Orton, 'An Approach to *Wulf and Eadwacer*', *Proceedings of the Royal Irish Academy*, 85 (1985), 223–58 (p. 225).
171. T. M. Charles-Edwards, 'The Social Background to Irish *Peregrinatio*', *Celtica*, 11 (1976), 43–59 (p. 46).
172. *Togail Bruidne Dá Derga: The Destruction of Dá Derga's Hostel*, ed. and trans. by Whitley Stokes (Paris: Bouillon, 1902), pp. 21–2.
173. Kim R. McCone, 'Werewolves, Cyclopes, Díberga, and Fíanna: Juvenile Delinquency in Early Ireland', *Cambridge Medieval Celtic Studies*, 12 (1986), 1–22 (p. 16).
174. These laws were first written down in the twelfth century, having supposedly been told to William the Conqueror by the English following his acquisition of the country – a tactic to legitimise William's rule; Bruce R. O'Brien, ed. and

trans., *God's Peace and King's Peace: The Laws of Edward the Confessor* (Philadelphia: University of Pennsylvania Press, 1999), p. 3. However, it is possible that they do include some of the laws that had governed England before the Norman invasion; see Marshall, *Wolves in Beowulf*, p. 28 n. 53.

175. *The Laws of Edward the Confessor*, ed. and trans. by O'Brien, p. 165. The outlaw continued to be termed a 'wolf's head' into the later medieval period, with a fourteenth-century Anglo-Norman legal treatise known as the *Mirror of Justices* describing how a person who did not appear in court when called upon on three occasions 'shall be accounted a wolf, and "Wolfshead!" shall be cried against him, for that a wolf is a beast hated of all folk; and from that time forward it is lawful for anyone to slay him like a wolf'. According to this text, the heads of wolves and outlaws alike were worth ten marks; *The Mirror of Justices*, ed. by William Joseph Whittaker (London: Quaritch, 1895), p. 125. Likewise, in a treatise written by thirteenth-century English jurist Henry de Bracton, it is said that outlaws 'bear the wolf's head and in consequence perish without judicial inquiry'; *Bracton on the Laws and Customs of England*, trans. by Samuel E. Thorne, 2 vols (Cambridge, MA: Belknap Press, 1968), II, p. 354.
176. Sarah L. Higley, 'Finding the Man Under the Skin: Identity, Monstrosity, Expulsion, and the Werewolf', in *The Shadow-walkers: Jacob Grimm's Mythology of the Monstrous*, ed. by T. A. Shippey (Tempe: Arizona Center for Medieval and Renaissance Studies, 2005), pp. 335–78 (p. 339). Emphasis in the original.
177. Hugo Edward Britt, 'The Beasts of Battle: Associative Connections of the Wolf, Raven and Eagle in Old English Poetry' (unpublished doctoral thesis, University of Melbourne, 2014), p. 138.
178. Translations my own.
179. John McKinnell, 'Eddic Poetry in Anglo-Scandinavian Northern England' in *Vikings and the Danelaw: Select Papers from the Proceedings of the Thirteenth Viking Congress, Nottingham and York, 21st–30th August 1997*, ed. by James Graham-Campbell et al. (Oxford: Oxbow, 2001), pp. 327–44 (p. 330).
180. Judith Jesch, 'The Norse Gods in England and the Isle of Man', in *Myths, Legends, and Heroes: Essays on Old Norse and Old English Literature in Honour of John McKinnell*, ed. by Daniel Anlezark (Toronto: University of Toronto Press, 2011), pp. 11–24 (p. 12).

Chapter 4: From Wolf Land to Wool Land

1. Jones, *The Medieval Natural World*, p. 78.
2. Contrary to the meaning of the term 'forest' today, royal Forests comprised not only woodland but all landscapes where deer might live, including marsh, heath, and grassland.
3. William's son, William II (r.1087–1100), introduced hunting rights for the nobility. Known as rights of 'chase', this gave landowners exclusive hunting rights over deer on their land, 'effectively remov[ing] the ancient hunting rights previously held by the landowner's tenants', which had allowed them to hunt and trap all wild animals on the land they occupied; Emma Griffin, *Blood Sport: Hunting in Britain Since 1066* (New Haven, CT: Yale University Press, 2007), p. 19.
4. Ibid., pp. 15–18.
5. Dominique Battles, *Cultural Difference and Material Culture in Middle English Romance: Normans and Saxons* (New York, NY: Routledge, 2013), p. 85.
6. Griffin, *Blood Sport*, p. 63.

7. Pluskowski, *Wolves and the Wilderness*, pp. 60, 89, and 75, and Battles, *Cultural Difference and Material Culture*, pp. 86 and 101.
8. William Perry Marvin, *Hunting Law and Ritual in Medieval English Literature* (Cambridge: Brewer, 2006), p. 51, translating from *The Peterborough Chronicle, 1070–1154*, ed. by Cecily Clark (Oxford: Oxford University Press, 1958), p. 13.
9. O'Connor, 'British Fauna in a Changing World', p. 7.
10. Pluskowski, *Wolves and the Wilderness*, p. 195.
11. Rowley, *Landscapes of the Norman Conquest*, p. 91 ('Map of Wales and the Welsh Marches showing land under Norman control in the first half of the twelfth century and areas of Flemish settlement').
12. William Linnard, 'The History of Forests and Forestry in Wales up to the Formation of the Forestry Commission' (unpublished doctoral thesis, Bangor University, 1979), pp. 79–80.
13. Richard Oram, *David I: The King Who Made Scotland* (Stroud: Tempus, 2004), p. 58, and John M. Gilbert, *Hunting and Hunting Reserves in Medieval Scotland* (Edinburgh: John Donald Publishers, 1979), p. 12.
14. Gilbert, *Hunting and Hunting Reserves in Medieval Scotland*, p. 13.
15. Pluskowski, *Wolves and the Wilderness*, p. 89.
16. Downham, *Medieval Ireland*, p. 190.
17. Fiona Beglane, *Anglo-Norman Parks in Medieval Ireland* (Dublin: Four Courts Press, 2015), pp. 60–1 and 56.
18. See Fiona Beglane, 'Forests and Chases in Medieval Ireland, 1169–*c.*1399', *Journal of Historical Geography*, 59 (2018), 90–9 (pp. 91 and 92–5).
19. Pluskowski, *Wolves and the Wilderness*, p. 195.
20. Yalden, *History of British Mammals*, p. 163.
21. Pluskowski, *Wolves and the Wilderness*, pp. 97–101, and Griffin, *Blood Sport*, p. 14.
22. Griffin, *Blood Sport*, p. 14.
23. Ibid., p. 21.
24. Thomas Blount, *Tenures of Land and Customs of Manors*, ed. by W. Carew Hazlitt, new edn (London: Reeves and Turner, 1874), p. 32. While we may associate the term 'wolf-dogs' with wolf–dog hybrids today, in these records it refers to hounds used to hunt wolves.
25. Blount, *Tenures of Land and Customs of Manors*, p. 219.
26. Ibid., p. 80.
27. Janet E. Burton, *The Foundation History of the Abbeys of Byland and Jervaulx* (York: Borthwick Publications, 2006), p. 53.
28. George R. Jesse, *Researches into the History of the British Dog*, 2 vols (London: Hardwicke, 1866), II, p. 17, citing *The Great Rolls of the Pipe for the Second, Third, and Fourth Years of the Reign of King Henry the Second, A.D. 1155, 1156, 1157, 1158*, ed. by Joseph Hunter (London: Eyre and Spottiswoode, 1844), pp. 54, 139, and 153.
29. Lee Raye, 'Wolves and Other Mammals Hunted in Medieval English Forests', in *The Wolf: Culture, Nature, Heritage*, ed. by Convery et al., pp. 37–46 (pp. 40–2).
30. Ibid., p. 40.
31. *Calendar of the Patent Rolls Preserved in the Public Record Office: Edward I, A.D. 1272–1281* (London: Eyre and Spottiswoode, 1901), p. 435.
32. Jesse, *Researches into the History of the British Dog*, p. 40.
33. *Southey's Common-place Book*, ed. by John Wood Warter (London: Longman, Brown, Green, and Longmans, 1849), p. 466.
34. Raye, 'Wolves and Other Mammals', p. 38.
35. *Regesta Regum Anglo-Normannorum, 1066–1154*, 4 vols (Oxford: Clarendon, 1913–69), II: *Regesta Henrici Primi*, ed. by Charles Johnson and H. A. Cronne (1956), p. 110.

36. Scottish king William I (r.1165–1214) was forced to transfer control of several castles to the English in 1174, including Edinburgh Castle, after a failed rebellion.
37. *Early Yorkshire Charters, Volume 9: The Stuteville Fee*, ed. by Charles Travis Clay (Cambridge: Cambridge University Press, 2013), pp. 3–4.
38. John Hudson, *The Oxford History of the Laws of England, Volume II: 871–1216* (Oxford University, 2012), p. 465 n. 77.
39. *Calendar of the Patent Rolls: Edward I, A.D. 1272–1281*, p. 429.
40. Ibid., p. 461.
41. *Calendar of the Patent Rolls Preserved in the Public Record Office: Edward III, A.D. 1370–1374* (London: Hereford Times, 1914), p. 135.
42. Richard Almond, *Medieval Hunting* (Stroud: The History Press, 2011), p. 70.
43. John Cummins, *The Hound and the Hawk: The Art of Medieval Hunting* (New York, NY: St Martin's Press, 1988), p. 138.
44. Raye, 'Wolves and Other Mammals', p. 39.
45. William's Forest Laws were unprecedented, but later documents such as this were forged to give legitimacy to the actions of his descendants. These documents made it seem as though the Normans were simply exercising their right to maintain Forests which, so the forged 'Forest Laws of Cnut' (one of the last pre-Norman kings) purport, had existed prior to the invasion; Marvin, *Hunting Law and Ritual in Medieval English Literature*, pp. 72–3.
46. Dolly Jørgensen, 'The Roots of the English Royal Forest', in *Anglo-Norman Studies XXXII: Proceedings of the Battle Conference 2009*, ed. by C. P. Lewis (Woodbridge: Boydell & Brewer, 2010), pp. 114–28 (p. 117).
47. 'Pseudo-Cnut De foresta', in *Die Gesetze der Angelsachsen*, ed. by Felix Liebermann, 3 vols (Halle: Niemeyer, 1903–16), i: *Text und Übersetzung* (1903), pp. 620–6 (p. 625).
48. Charles McLean Andrews, *The Old English Manor: A Study in English Economic History* (Baltimore: John Hopkins, 1892), p. 227. See Marvin, *Hunting Law and Ritual in Medieval English Literature*, p. 73 for the details of some of these punishments.
49. Pluskowski, *Wolves and the Wilderness*, p. 104.
50. *Welsh Medieval Law: Being a Text of the Laws of Howel the Good*, ed. and trans. by A. W. Wade-Evans (Oxford: Clarendon, 1909), p. 225.
51. Yalden, *The History of British Mammals*, p. 168, and Austin Lane Poole, *From Domesday Book to Magna Carta, 1087–1216* (Oxford: Oxford University Press, 1993), p. 31.
52. '1428/3/6', *The Records of the Parliaments of Scotland to 1707*, ed. by Keith M. Brown et al. (University of St Andrews, 2007–23) <https://www.rps.ac.uk/trans/1428/3/6> [accessed 10 July 2023].
53. '1458/3/36', *The Records of the Parliaments of Scotland to 1707*, ed. by Brown et al. <https://www.rps.ac.uk/trans/1458/3/36> [accessed 10 July 2023].
54. *Calendar of the Patent Rolls: Edward I, A.D. 1272–1281*, p. 436.
55. See p. 102 above.
56. Pluskowski, *Wolves and the Wilderness*, p. 29.
57. 'Parishes: Mathon', in *A History of the County of Worcester*, ed. by William Page and J. W. Willis-Bund, 5 vols (London: Constable and Co., 1901–26), iv (1924), pp. 139–43. Available at: *British History Online* <http://www.british-history.ac.uk/vch/worcs/vol4/> [accessed 5 May 2023].
58. Pluskowski, *Wolves and the Wilderness*, p. 28, and Rackham, *History of the Countryside*, p. 35.
59. *Calendar of Ormond Deeds, 1172–1350*, ed. by Edmund Curtis (Dublin: Stationery Office, 1932), pp. 3–4.

60. Pluskowski, 'Where Are the Wolves?', p. 291.
61. Peter Yeoman, *Medieval Scotland: An Archaeological Perspective* (London: Batsford, 1995), p. 94.
62. Pluskowski, *Wolves and the Wilderness*, pp. 22–3. Anecdotes of wolf remains found at Walthamstow (London) and Pevensey (Sussex) also exist, but the physical evidence is long gone; ibid.
63. Umberto Albarella, 'A Review of Animal Bone Evidence from Central England', *Historic England Research Report Series*, 61 (2019), p. 206, and Simon Davis, 'Animal Bone from Langham Road and Burystead', in *Raunds: The Origin and Growth of a Midland Village, AD 450–1500 : Excavations in North Raunds, Northamptonshire 1977–87*, ed. by Michel Audouy and Andy Chapman (Oxford: Oxbow, 2009), pp. 214–21 (p. 217).
64. Karen Lloyd, 'The Helsfell Wolf', in *The Wolf: Culture, Nature, Heritage*, ed. by Convery et al., pp. 371–6 (p. 375).
65. Pluskowski, *Wolves and the Wilderness*, p. 30.
66. Mark Brown, 'Wellington's False Teeth and Wolf Bones: English Heritage Seeks Help to Record Vast Collection', *Guardian*, 8 October 2024 <https://www.theguardian.com/uk-news/2024/oct/08/wellingtons-false-teeth-and-wolf-bones-english-heritage-needs-help-to-record-vast-collection> [accessed 18 December 2024].
67. Edward, Second Duke of York, *The Master of Game: The Oldest English Book on Hunting*, ed. by Wm. A. and F. Baillie-Grohman (London: Chatto and Windus, 1909), p. 63.
68. Meaney, *Anglo-Saxon Amulets and Curing Stones*, p. 57, and Robert Max Garrett, 'Middle English Rimed Medical Treatise', *Anglia*, 34 (1911), 163–93 (pp. 168–9, lines 172–88).
69. Gerald of Wales, *The Journey through Wales*, in *The Journey through Wales and The Description of Wales*, trans. by Lewis Thorpe (Harmondsworth: Penguin, 1978), pp. 63–209 (p. 130).
70. Aleksander G. Pluskowski, 'The Wolf', in *Extinctions and Invasions*, ed. by O'Connor and Sykes, pp. 68–74 (p. 73).
71. Edward, Second Duke of York, *The Master of Game*, p. 63.
72. *Ancient Laws and Institutes of Wales*, ed. by Aneurin Owen (London: Eyre and Spottiswoode, 1841), p. 141.
73. Gilbert, *Hunting and Hunting Reserves in Medieval Scotland*, p. 213.
74. Pluskowski, *Wolves and the Wilderness*, pp. 113–14.
75. Pluskowski, 'Where Are the Wolves?', p. 291.
76. Rackham, *History of the Countryside*, p. 35, and Pluskowski, 'The Wolf', p. 73.
77. Faiers, *Fur: A Sensitive History*, pp. 89 and 100.
78. Pluskowski, *Wolves and the Wilderness*, p. 91.
79. Salisbury, *The Beast Within*, p. 25.
80. Pluskowski, 'Wolves and Sheep in Medieval Semiotics', pp. 16–17.
81. Pluskowski, *Wolves and the Wilderness*, p. 91.
82. Edward, Second Duke of York, *The Master of Game*, p. 61.
83. Adrian R. Bell, Chris Brooks, and Paul R. Dryburgh, *The English Wool Market, c.1230–1327* (Cambridge: Cambridge University Press, 2007), p. 1. Emphasis in the original.
84. Ibid., p. 9.
85. Pluskowski, 'Wolves and Sheep in Medieval Semiotics', pp. 14–15.
86. Winder, *The Last Wolf*, p. 199.
87. John T. Maple, 'Anglo-Norman Conquest of Ireland and the Irish Economy: Stagnation or Stimulation?', *The Historian*, 52 (1989), 61–81 (p. 73).

88. Timothy O'Neill, 'Trade', in *Medieval Ireland: An Encyclopedia*, ed. by Duffy, pp. 449–51 (p. 450).
89. T. C. Smout, Alan R. MacDonald, and Fiona Watson, *A History of the Native Woodlands of Scotland, 1500–1920* (Edinburgh: Edinburgh University Press, 2007), p. 40.
90. J. Geraint Jenkins, *The Welsh Woollen Industry* (Cardiff: Lewis, 1969), p. 98.
91. D. Huw Owen, 'Wales and the Marches', in *The Agrarian History of England and Wales, Volume 3: 1348–1500*, ed. by Edward Miller (Cambridge: Cambridge University Press, 1991), pp. 238–54 (p. 244).
92. Jenkins, *The Welsh Woollen Industry*, p. 99.
93. David Stone, 'The Productivity and Management of Sheep in Late Medieval England', *Agricultural History Review*, 51.1 (2003), 1–22 (p. 1).
94. Ibid., p. 2.
95. Linnard, 'The History of Forests and Forestry in Wales', p. 109.
96. Ian D. Whyte, *Scotland Before the Industrial Revolution: An Economic and Social History c.1050–c.1750* (New York and London: Routledge, 1995), p. 45, and Geraldine Stout, 'The Cistercian Grange: A Medieval Farming System', in *Agriculture and Settlement in Ireland*, ed. by Margaret Murphy and Matthew Stout (Four Courts Press, 2015), pp. 28–68 (p. 28).
97. Linnard, 'The History of Forests and Forestry in Wales', p. 109, and R. R. Davies, *The Age of Conquest: Wales, 1063–1415* (Oxford: Oxford University Press, 2000), p. 170.
98. Robert Trow-Smith, *A History of British Livestock Husbandry, to 1700* (Abingdon: Routledge, 2006), p. 139.
99. Winder, *The Last Wolf*, p. 75.
100. Ibid., p. 81–2 and 84.
101. Ibid., p. 4.
102. Paul Williams, *Howls of Imagination: Wolves of England* (Loughborough: Heart of Albion, 2007), p. 8.
103. Dent, *Lost Beasts of Britain*, p. 114, and Anthony Dent, 'The Last of the Wolf', *History Today*, 24.2 (1974), 120–7 (p. 123).
104. Winder, *The Last Wolf*, p. 33.
105. Ibid., pp. 33 and 22 respectively.
106. Coulthard, *A Short History of the World According to Sheep*, p. 193.
107. Dent, *Lost Beasts of Britain*, p. 114.
108. Aybes and Yalden, 'Place-name Evidence', p. 202. While the transformation of wolf place-names into wool place-names is the result of linguistic mutation, it aptly reflects the transformation of Wolf Land to Wool Land.
109. Rackham, *History of the Countryside*, p. 88.
110. Ibid., p. 110.
111. Ibid., p. 109.
112. Pluskowski, 'The Tyranny of the Gingerbread House', p. 153.
113. Ibid.
114. Pluskowski, 'Where Are the Wolves?', pp. 285–6.
115. Aleks Pluskowski, 'The Medieval Wild', in *The Oxford Handbook of Later Medieval Archaeology in Britain*, ed. by Christopher M. Gerrard and Alejandra Gutiérrez (Oxford: Oxford University Press, 2018), pp. 141–53 (p. 145).
116. Ibid., p. 141.
117. Nicholas J. Mayhew, 'Scotland: Economy and Society', in *A Companion to Britain in the Later Middle Ages*, ed. by S. H. Rigby (Malden: Blackwell, 2008, pp. 107–24 (p. 109).
118. H. B. Clarke, 'Population', in *Medieval Ireland: An Encyclopedia*, ed. by Duffy, pp. 383–4 (p. 384).

119. Downham, *Medieval Ireland*, p. 193, and O'Sullivan, 'Woodlands', p. 523.
120. Hickey, *Wolves in Ireland*, p. 64.
121. Sands, 'Dewilding "Wolf-land"', p. 262.
122. Pluskowski, *Wolves and the Wilderness*, p. 49.
123. Pluskowski, 'The Medieval Wild', p. 145.
124. Wiseman, 'A Noxious Pack', p. 105.
125. *Mediæval Lore from Bartholomaeus Anglicus*, ed. and trans. by Robert Steele (London: Moring, 1905), pp. 164–5.
126. Willene B. Clark, *A Medieval Book of Beasts: The Second Family Bestiary. Commentary, Art, Text and Translation* (Woodbridge: Boydell, 2006), pp. 142–4.
127. Wilma B. George and William Brunsdon Yapp, *The Naming of the Beasts: Natural History in the Medieval Bestiary* (London: Duckworth, 1991), p. 51.
128. Donalson, *The History of the Wolf in Western Civilization*, p. 100.
129. Such was their popularity that an illustrated collection of Aesop's fables, produced in 1484 by William Caxton (c.1422–c.1491), was one of the earliest books printed in Britain.
130. Pluskowski, *Wolves and the Wilderness*, p. 123.
131. 'The Fox and Wolf in the Well', in *Middle English Humorous Tales in Verse*, ed. by George H. McKnight (Boston, MA and London: Heath & Co., 1913), pp. 25–37 (p. 36, line 282). Spelling modernised.
132. See Gerald of Wales, *The History and Topography of Ireland*, trans. by John J. O'Meara (London: Penguin, 1982), pp. 69–72.
133. Luuk Houwen, 'Howling Wolves and Other Beasts Animals and Monstrosity in the Middle Ages', in *Animals, Animality, and Literature*, ed. by Bruce Boehrer, Molly Hand, and Brian Massumi (Cambridge: Cambridge University Press, 2018), pp. 43–56 (p. 49).
134. John D. C. Linnell and Julien Alleau, 'Predators That Kill Humans: Myth, Reality, Context and the Politics of Wolf Attacks on People', in *Problematic Wildlife: A Cross-Disciplinary Approach*, ed. by Francesco M. Angelici (Cham: Springer, 2016), pp. 357–71 (pp. 364–5).
135. *Mediæval Lore from Bartholomaeus Anglicus*, ed. and trans. by Steele, p. 164.
136. Edward, Second Duke of York, *The Master of Game*, pp. 59–61.
137. Linnell et al., 'The Fear of Wolves', p. 36.
138. Pluskowski, *Wolves and the Wilderness*, p. 31.
139. Gervase of Tilbury, *Otia Imperialia, Recreation for an Emperor*, ed. and trans. by S. E. Banks and J. W. Binns (Oxford: Clarendon, 2002), p. 813.
140. Beresford, *The White Devil*, pp. 170–1.
141. Marvin, *Wolf*, p. 53.
142. Annal 1420.7, in *Annála Connacht*, comp. by Pádraig Bambury (CELT: Corpus of Electronic Texts: 2008) <https://celt.ucc.ie/published/T100011/index.html> [accessed 06 January 2025]; Mac Coitir, *Ireland's Animals*, p. 51; and Kelly, *Early Irish Farming*, p. 187.
143. Annal T1137.8, in *The Annals of Tigernach*, trans. by Gearóid Mac Niocaill (CELT: Corpus of Electronic Texts, 2010) <https://celt.ucc.ie/published/T100002A/index.html> [accessed 06 January 2025].
144. Walter Map, *De Nugis Curialium: Courtiers' Trifles*, ed. and trans. by M. R. James, rev. by C. N. L. Brooke and R. A. B. Mynors (Oxford: Clarendon, 1983), p. 185.
145. See Marshall, 'A "Wasteland" Infested by Wolves'.
146. William Hutchinson, *The History of the County of Cumberland*, 2 vols (Carlise: Jollie, 1794), II, p. 16.
147. Jennifer Westwood and Jacqueline Simpson, *The Lore of the Land: A Guide to England's Legends, from Spring-heeled Jack to the Witches of Warboys* (London: Penguin, 2005), p. 125.

148. Edward, Second Duke of York, *The Master of Game*, p. 60.
149. Giolla Críost Táilliúr, 'Beannuigh do Theaghlach, a Thríonóid', in *Duanaire na Sracaire/Songbook of the Pillagers: Anthology of Medieval Gaelic Poetry*, ed. by Wilson McLeod and Meg Bateman (Edinburgh: Birlinn, 2007), pp. 234–9, cited in Wiseman, 'A Noxious Pack', pp. 106–9.
150. Pluskowski, 'The Tyranny of the Gingerbread House', p. 147.
151. Pluskowski, *Wolves and the Wilderness*, p. 150.
152. Although the Anglo-Norman invaders brought heraldry to Ireland, it is not clear whether the practice was adopted by the native Irish before the seventeenth century; John Barry, 'Guide to Records of the Genealogical Office, Dublin, with a Commentary on Heraldry in Ireland and on the History of the Office', *Analecta Hibernica*, 26 (1970), 3–43.
153. Pluskowski, *Wolves and the Wilderness*, p. 134, and Aleks Pluskowski, 'The Zooarchaeology of Medieval "Christendom": Ideology, the Treatment of Animals and the Making of Medieval Europe', *World Archaeology*, 42 (2010), 201–14 (p. 206).
154. This coat of arms has been digitised on <http://www.aspilogia.com/HE-Heralds_Roll/N-145-216.html> [accessed 05 February 2025].
155. Beresford, *The White Devil*, p. 12; John W. Papworth, *An Alphabetical Dictionary of Coats of Arms Belonging to Families in Great Britain and Ireland*, 2 vols (London: Richards, 1874), ii, p. 912; and Pluskowski, *Wolves and the Wilderness*, p. 149.
156. Wilhelmina Powlett, Duchess of Cleveland, *The Battle Abbey Roll*, 3 vols (London: Murray, 1889), ii, pp. 188–91.
157. Wiseman, 'A Noxious Pack', p. 106.
158. Poole, *From Domesday Book to Magna Carta*, p. 31 n. 3.
159. Although wolves were exterminated because they competed with people for deer, this did little to stop the deer population from becoming severely depleted by the fifteenth century thanks to over-hunting and poor management of Forests; Pluskowski, 'The Wolf', p. 74.
160. Pluskowski, 'The Medieval Wild', p. 141.
161. Ranulf Higden, *Polychronicon*, trans. by John Trevisa and ed. by Joseph Rawson Lumby, 9 vols (London: Longman, Green, Longman, Roberts, and Green, 1865–86), ii (1869), p. 17.
162. The man claimed that he needed the wolves to treat a disease called 'le lou', but when questioned he admitted that neither he nor anyone else he knew suffered from this illness, nor was he even a doctor. The physicians and surgeons of London were summoned to the court, and testified that they knew of no disease for which wolf flesh could be used as treatment; Faye Getz, *Medicine in the English Middle Ages* (Princeton, NJ: Princeton University Press, 1998), p. 76.
163. Cummins, *The Hound and the Hawk*, p. 137.
164. Edward, Second Duke of York, *The Master of Game*, p. 54. Emphasis in the original.
165. Raye, 'Wolves and Other Mammals', p. 43.
166. *The Remains of John Briggs* (Kirkby Lonsdale: Foster, 1825), pp. 35–7, and Tristram [pseudonym], 'The Last Wolf. A Legend of Humphrey Head', in *Fugitive Poems* (London: Hardwicke, 1855), pp. 9–27.
167. This content of the tale, from both the letter and poem, is summarised in Edwin Waugh's *Rambles in the Lake Country and Its Borders* (Manchester: Heywood, 1864), pp. 77–83.
168. *The Remains of John Briggs*, pp. 36–7.
169. Dent, *Lost Beasts of Britain*, pp. 131–2.
170. See Fychan, *Galwad y Blaidd*, pp. 68–87, and John Pollard, *Wolves and Werewolves* (London: Hale, 1964), pp. 76–84.

171. Westwood and Simpson, *Lore of the Land*, p. 137.
172. Ibid., p. 539.
173. Almond, *Medieval Hunting*, p. 71.
174. Dent, *Lost Beasts of Britain*, p. 126.

Chapter 5: 'Enlightenment': The Final Purge

1. Boece, *Scotorum Historia*, ed. and trans. by Sutton, Preliminary Matter <https://philological.cal.bham.ac.uk/boece/fronteng.html> [accessed 13 January 2025].
2. John Lesley, *The Historie of Scotland*, trans. by James Dalrymple and ed. by E. G. Cody and William Murison, 2 vols (Edinburgh and London: Blackwood and Sons, 1888–95), I (1888), ed. by E. G. Cody (p. 29). Spelling modernised.
3. *Holinshed's Chronicles of England, Scotland, and Ireland*, V: *Scotland* (1808), p. 13. Spelling modernised.
4. John Taylor, *The Pennyles Pilgrimage* (London: the author, 1618), p. 50.
5. See Rob Lenders, '"Wild" Horses in Medieval and Early Modern Landscapes of Europe', *Landscape History*, 44.2 (2023), 15–35.
6. John G. Harrison, 'The Last Wolf in Scotland – Some Historical Evidence' (2020), p. 6 <https://www.johnscothist.com/uploads/5/0/2/4/5024620/harrison_last_wolf_v._4.pdf> [accessed 15 September 2021].
7. Wiseman, 'A Noxious Pack', p. 117, citing Mark Louden Anderson, *A History of Scottish Forestry*, ed. by Charles J. Taylor, 2 vols (London: Nelson, 1967), I: *From the Ice Age to the French Revolution*, p. 275.
8. Amy L. Juhala, 'The Household and Court of King James VI of Scotland, 1567–1603' (unpublished doctoral thesis, University of Edinburgh, 2000), p. 361.
9. *Accounts of the Lord High Treasurer of Scotland*, ed. by James Balfour Paul, Thomas Dickson, and Charles Thorpe McInnes, 13 vols (Edinburgh: H. M. General Register House, 1877–1978), III: A.D. 1506–1507, ed. by Sir James Balfour Paul (1901), pp. 170, xxiii, and lvii.
10. Lesley, *The Historie of Scotland*, I, p. 7. Spelling modernised.
11. John Stewart-Murray, Seventh Duke of Atholl, *Chronicles of the Atholl and Tullibardine Families*, 5 vols (Edinburgh: Ballantyne Press, 1908), I, pp. 36–7, cited in Wiseman, 'A Noxious Pack', pp. 119–20.
12. *Rental Book of the Cistercian Abbey of Cupar-Angus*, ed. by Charles Rogers, 2 vols (London: Grampian Club, 1879–80), II (1880), p. 107.
13. *The Black Book of Taymouth with Papers from the Breadalbane Charter Room*, ed. by Cosmo Innes (Edinburgh: Bannatyne Club, 1855), p. 356.
14. The actual surname of the brothers was Allen, and they claimed to be the grandsons of Jacobite pretender Charles III. The pair wrote numerous dubious works on Scottish history and culture, including an infamous compendium of clan tartans which, though they claimed to have sourced from a fifteenth-century manuscript, was forged.
15. John Sobieski and Charles Edward Stuart, *Lays of the Deer Forest*, 2 vols (Edinburgh and London: Blackwood and Sons, 1848), II, pp. 244–5, 229, and 231–2.
16. Harting, *British Animals*, p. 168.
17. See pp. 161–2 below.
18. Smout, MacDonald, and Watson, *A History of the Native Woodlands of Scotland*, p. 20.
19. Richard Franck, 'Franck's Memoirs', in *Early Travellers in Scotland*, ed. by P. Hume Brown (Edinburgh: David Douglas, 1891), pp. 184–216 (p. 202). Spelling modernised.

20. Thomas Kirke, 'A Modern Account of Scotland by an English Gentleman (1679)', in *Early Travellers in Scotland*, ed. by Brown, pp. 251–65 (p. 254).
21. Mairi Stewart, 'Using the Woods, 1600-1850: (1) The Community Resource', in *People and Woods in Scotland: A History*, ed. by T. C. Smout (Edinburgh: Edinburgh University Press, 2003), pp. 82–104 (p. 83).
22. Ian D. Whyte, *Landscape and History Since 1500* (London: Reaktion, 2004), p. 49.
23. Wiseman, 'A Noxious Pack', p. 131.
24. Ibid., p. 104.
25. Smout, MacDonald, and Watson, *A History of the Native Woodlands of Scotland*, pp. 37–9.
26. Ibid., p. 45.
27. Ibid., p. 42.
28. See p. 64 above.
29. Smout, MacDonald, and Watson, *A History of the Native Woodlands of Scotland*, p. 42.
30. Crumley, *The Last Wolf*, pp. 57–8.
31. Wiseman, 'A Noxious Pack', p. 131.
32. Ibid., p. 114.
33. Ibid., p. 131.
34. Ibid., pp. 131–2.
35. MS Pont 3, National Library of Scotland <https://maps.nls.uk/counties/rec/258> [accessed 13 September 2023]. Spelling modernised.
36. William Camden, 'Scotland', in *Britain, or a Chorographical Description of the Most Flourishing Kingdoms, England, Scotland, and Ireland*, trans. by Philemon Holland (London: Bishop and Norton, 1610), pp. 1–54 (p. 54). Spelling modernised.
37. *Geographical Collections Relating to Scotland Made by Walter Macfarlane*, ed. by Arthur Mitchell and James Toshach Clark, 3 vols (Edinburgh: Constable, 1906–8), II, ed. by Arthur Mitchell (1907), p. 454.
38. This tradition was the subject of an oft-cited poem written by Eliza Ogilvy, which was published in her *A Book of Highland Minstrelsy* (London: Bosworth, 1848), pp. 251–9.
39. Wiseman, 'A Noxious Pack', pp. 116–17. Similar folklore also comes from Ireland. For burials at the Gate Cemetery in Ogonnelloe, Co. Clare, it is apparently customary to circle the cemetery walls with the coffin and place it at various points on the ground, a practice begun in the 1930s but said to have originated from past efforts to prevent wolves from knowing the true location of the burial. Elsewhere in Co. Clare, it was supposedly customary to build two coffins, leaving one empty to fool both wolves and the devil; Hickey, *Wolves in Ireland*, p. 48.
40. Crumley, *The Last Wolf*, p. 64.
41. Almond, *Medieval Hunting*, p. 71.
42. Robert Sibbald, *The Wild Plants of Scotland and the Animals of Scotland*, trans. by Lee Raye (Cardiff: KDP, 2020), p. 270.
43. Thomas Pennant, *History of Quadrupeds*, 2 vols (London: White, 1781), I, p. 231.
44. In 'A Noxious Pack' (p. 126), Wiseman provides a long list of the places at which various 'last wolves' in Scotland are claimed to have been killed.
45. William Scrope, *The Art of Deer-stalking* (London: Arnold, 1897), pp. 274–7.
46. In 2019, artist Beatrice Searle laid a new stone nearby, which reads: 'In memory of the wolves / Part of these lands / Lost to generations / We await your return'; 'Anna Souter & Beatrice Searle in Conversation', *Assemblage*, April 2019

<https://assemblagemag.wixsite.com/assemblage-magazine/anna-souter-beatrice-searle> [accessed 19 November 2025].

47. James Hogg, *Winter Evening Tales, Collected Among the Cottagers in the South of Scotland* (New York: Leavitt & Allen Bros., 1873), pp. 170–2.
48. Scrope, *The Art of Deer-stalking*, p. 277 n. 1.
49. Pollard, *Wolves and Werewolves*, pp. 82–3, 90, and 92.
50. Scrope, *The Art of Deer-stalking*, p. 277 n. 1.
51. Wiseman, 'A Noxious Pack', p. 131.
52. Harting, *British Animals*, pp. 174–5. See also Gow, *Hunt for the Shadow Wolf*, p. 91, Crumley, *The Last Wolf*, p. 76, and Wiseman, 'A Noxious Pack', pp. 126–7 and 128.
53. Wiseman, 'A Noxious Pack', p. 127.
54. Crumley, *The Last Wolf*, p. 17. This story was told by Otta F. Swire in her 1952 collection of folkloric tales from Skye; *Skye: The Island and Its Legends* (London: Oxford University Press, 1952), pp. 230–1. However, other records note that the clan's coat of arms depicts a boar with a bone in its mouth; *Elven's Heraldry* (London: Barfield, 1815), plate 15, no. 24.
55. Thomas Pennant, *A Tour in Scotland, 1769* (Edinburgh: Birlinn, 2000), p. 138 n. 13.
56. Harting, *British Animals*, p. 175.
57. Crumley, *The Last Wolf*, p. 74.
58. This book, Weymouth noticed, includes an illustration of the wolf which looks very similar to one in Donovan's *Natural History of British Quadrupeds*, which he assumes was inspired by the same animal; Adam Weymouth, 'Was This the Last Wild Wolf of Britain?', *Guardian*, 21 July 2014 <https://www.theguardian.com/science/animal-magic/2014/jul/21/last-wolf> [accessed 10 September 2020].
59. Ibid.
60. Kitchener, 'Extinctions, Introductions and Colonisations of Scottish Mammals and Birds Since the Last Ice Age', p. 75.
61. Versions of this account are told by Scottish author Sir Thomas Dick Lauder, rather bizarrely as part of his account of *The Great Floods of August 1829 in the Province of Moray, and Adjoining Districts*, second edn (Edinburgh: Black, 1830), pp. 44–6), and by Sobieski and Stuart in *Lays of the Deer Forest*, II, pp. 244–7, both of which are quoted from here. Lauder also wrote a novel about the fourteenth-century Earl of Buchan, Alexander Stewart (who was, incidentally, known as the 'Wolf of Badenoch'), in which Lauder depicts a knight fighting an 'enormous' monstrous wolf with 'oblique and sinister eyes […] flashing fire', 'frothy jaws' as well as 'long sharp tusks'; Sir Thomas Dick Lauder, *The Wolfe of Badenoch: A Historical Romance of the Fourteenth Century* (London: Hamilton, Adams and Co., 1886), pp. 91–2.
62. Wiseman, 'A Noxious Pack', p. 130, citing *Geographical Collections Relating to Scotland Made by Walter Macfarlane*, ed. by Mitchell and Clark, II, ed. by Mitchell (1907), p. 270.
63. Crumley, *The Last Wolf*, p. 4.
64. Derek Gow cites the name of the Northamptonshire village of Blakesley as evidence for the presence of black wolves in Britain (*Hunt for the Shadow Wolf*, p. 66), but this name, which means 'black wolf's meadow', does not pertain to an animal but rather to a person named Blæcwulf, to whom the meadow belonged; A. D. Mills, *A Dictionary of British Place-names* (Oxford: Oxford University Press, 2003), p. 61.

    The gene mutation that creates a black coat in wolves came about through hybridisation with domestic dogs; Elizabeth Pennisi, 'Borrowed Gene Blackens Wolves: Interbreeding with Dogs Gave Wolves an Evolutionary Advantage',

Science, 5 February 2009 <www.science.org/doi/10.1126/article.31377> [accessed 10 January 2025] and Małgorzata Pilot et al., 'Widespread, Long-term Admixture Between Grey Wolves and Domestic Dogs Across Eurasia and Its Implications for the Conservation Status of Hybrids', *Evolutionary Applications*, 11 (2018), 662–80 (p. 673). Black wolves are much more common in North America than they are in Europe, however, the explanation for which appears to lie with a deadly disease known as canine distemper virus (CDV). The gene that produces a black coat also helps to protect against this disease, and black wolves are more common in places where CDV has occurred; Sarah Cubaynes et al., 'Disease Outbreaks Select for Mate Choice and Coat Color in Wolves', *Science*, 378 (2022) <www.science.org/doi/10.1126/science.abi8745>. CDV originated in the Americas when the measles virus, brought over by Europeans, crossed into dogs, with the first case recorded in South America in 1735; Rebecca P. Wilkes, 'Canine Distemper Virus in Endangered Species: Species Jump, Clinical Variations, and Vaccination', *Pathogens*, 12.1 (2022), 1–17 (pp. 1–2). CDV subsequently made its way to Europe in the 1760s (Elizabeth W. Uhl et al., 'New World Origin of Canine Distemper: Interdisciplinary Insights', *International Journal of Paleopathology*, 24 (2019), 266–78 (p. 269, Table 1), several decades after MacQueen's black wolf was supposedly ravaging the Findhorn.

The best defence for this element of the story is actually that the black wolf described in English retellings of the tale may be a mistranslation of the Gaelic *dubh*, which can mean 'black' but which also can also mean 'dark'; Pollard, *Wolves and Werewolves*, p. 88. But, all things considered, this is little defence for the truth of a story clearly steeped in local and family folklore and tradition.

65. Stefan Buczacki, *Fauna Britannica* (London: Hamlyn, 2002), p. 404.
66. Pollard, *Wolves and Werewolves*, p. 88. See Alexander Nisbet, *A System of Heraldry, Speculative and Practical* (Edinburgh: MackEuen, 1722), p. 331.
67. Crumley, *The Last Wolf*, p. 66.
68. Ironically, research has found that black wolves are less aggressive than their grey counterparts; Bridgett M. vonHoldt et al., 'Heritability of Interpack Aggression in a Wild Pedigreed Population of North American Grey Wolves', *Molecular Ecology*, 29 (2020), 1764–75 (p. 1765).
69. Crumley, *The Last Wolf*, pp. 100–1 and 103–4.
70. Ibid., p. 104.
71. Sands, 'Dewilding "Wolf-land"', p. 262.
72. Hickey, *Wolves in Ireland*, pp. 89–90.
73. Kieran R. Hickey, 'A Geographical Perspective on the Decline and Extermination of the Irish Wolf *Canis lupus*—An Initial Assessment', *Irish Geography*, 33 (2000), 185–98 (p. 194).
74. Patrick Sleeman, 'Mammals and Mammalogy', in *Nature in Ireland: A Scientific and Cultural History*, ed. by John Wilson Foster and Helena C. G. Chesney (Dublin: Lilliput Press, 1997), 241–61 (p. 245).
75. George Gascoigne, *The Noble Art of Venerie or Hunting* (London: Purfoot, 1611), p. 203.
76. Hickey, 'A Geographical Perspective on the Decline and Extermination of the Irish Wolf', p. 194.
77. Sands, 'Dewilding "Wolf-land"', p. 262.
78. G. L. Gomme, 'Totemism in Britain', *The Archaeological Review*, 3 (1889), 217–42 (p. 221 n. 4).
79. Edmund Spenser, *A View of the Present State of Ireland*, ed. by W. L. Renwick (Oxford: Clarendon, 1970), p. 59.

80. Camden, 'Ireland, and the Smaller Ilands in the British Ocean', in *Britain, or a Chorographical Description of the Most Flourishing Kingdoms, England, Scotland, and Ireland*, trans. by Holland, pp. 55–233 (p. 146). Spelling modernised.
81. Donna L. Potts, *Contemporary Irish Writing and Environmentalism: The Wearing of the Deep Green* (Cham: Palgrave Macmillan, 2018), pp. 14–15.
82. Thomas Blenerhasset, 'A Direction for the Plantation in Ulster', in *A Contemporary History of Affairs in Ireland, from 1641 to 1652, Vol. 1 Part 1*, ed. by John T. Gilbert (Dublin: For the Irish Archaeological and Celtic Society, 1879), pp. 317–26 (pp. 318 and 320).
83. Willy Maley, ed., 'The Supplication of the Blood of the English Most Lamentably Murdered in Ireland, Cryeng Out of the Yearth for Revenge (1598)', *Analecta Hibernica*, 36 (1995), 1 and 3–77 (pp. 12, 18, and 74), quoted in Eamon Darcy, *The Irish Rebellion of 1641 and the Wars of the Three Kingdoms* (Woodbridge: Boydell, 2013), p. 34. Spelling modernised.
84. Potts, *Contemporary Irish Writing and Environmentalism*, p. 147.
85. Hickey, *Wolves in Ireland*, p. 68. See *Irelands Tragical Tyrannie Sent Over in Two Letters, by a Speehlesse Damzell* (London: Printed for T. L., 1642) <https://quod.lib.umich.edu/cgi/t/text/text-idx?c=eebo;idno=A57457.0001.001> [accessed 13 September 2023].
86. Potts, *Contemporary Irish Writing and Environmentalism*, pp. 14–15.
87. Eileen McCracken, *The Irish Woods Since Tudor Times: Distribution and Exploitation* (Newton Abbot: David & Charles, 1971), p. 27.
88. Hickey, 'A Geographical Perspective on the Decline and Extermination of the Irish Wolf', pp. 188–90.
89. Fairley, *An Irish Beast Book*, p. 295, and Hickey, 'A Geographical Perspective on the Decline and Extermination of the Irish Wolf', p. 194.
90. Hickey, 'A Geographical Perspective on the Decline and Extermination of the Irish Wolf', p. 193.
91. Sands, 'Dewilding "Wolf-land"', p. 263.
92. Ibid.
93. George Bennett, *The History of Bandon and the Principal Towns in the West Riding of County Cork*, enlarged edn (Cork: Guy, 1869), pp. 92–3, cited in Hickey, *Wolves in Ireland*, p. 66.
94. Hickey, *Wolves in Ireland*, p. 102.
95. Philip O'Sullivan Beare, *Ireland Under Elizabeth: Chapters Towards a History of Ireland in the Reign of Elizabeth*, ed. and trans. by Matthew J. Byrne (Dublin: Sealy, Bryers & Walker, 1903), p. 181.
96. Sands, 'Dewilding "Wolf-land"', p. 262, citing Fynes Moryson, 'The Description of Ireland', in *Illustrations of Irish History and Topography, Mainly of the Seventeenth Century*, ed. by C. Litton Falkiner (London: Longmans, Green, and Co., 1904), pp. 214–32 (p. 222).
97. O'Sullivan Beare, *Ireland Under Elizabeth*, p. 181.
98. Hickey, *Wolves in Ireland*, p. 102.
99. These people may have learned how to hunt wolves in Scotland and Europe. Killing wolves was perhaps entrusted to English rather than Gaelic hunters because, as Kieran Hickey observes, 'politically, it may have been that the prospect of numbers of armed Irish roaming around the country hunting wolves was not acceptable, given the ongoing conflict between the Irish and the new English settlers and attempts to control the country and prevent further rebellions'; Hickey, 'A Geographical Perspective on the Decline and Extermination of the Irish Wolf', p. 195.
100. Ibid., p. 196.
101. Hickey, *Wolves in Ireland*, p. 94.

102. Sands, 'Dewilding "Wolf-land"', p. 263. An estimated 850,000 of the almost 1.5 million-strong Irish population died in just eleven years between 1641 and 1652, by which time 'the plague and famine had swept away whole counties', so that 'a man might travel twenty or thirty miles, and not see a living creature, either man, beast, or bird, they being either all dead, or had quit those desolate places', Thomas L. Coonan, *The Irish Catholic Confederacy and the Puritan Revolution* (Dublin: Clonmore & Reynolds, 1954), pp. 327–8, citing Richard Lawrence, *The Interest of Ireland in Its Trade and Wealth Stated* (Dublin: Ray and Howes, 1682), Part 2, pp. 86–7.
103. Hickey, *Wolves in Ireland*, p. 68, and Hickey, 'A Geographical Perspective on the Decline and Extermination of the Irish Wolf', p. 194.
104. *Ireland Under the Commonwealth: Being a Selection of Documents Relating to the Government of Ireland from 1651 to 1659*, ed. by Robert Dunlop, 2 vols (Manchester: Manchester University Press, 1913), II, p. 340, and Sleeman, 'Mammals and Mammalogy', p. 251.
105. Wolfhounds, sometimes referred to as the Irish greyhound, became very rare after the eradication of wolves from Ireland; Michael Viney, 'Wild Sports and Stone Guns', in *Nature in Ireland*, ed. by Foster and Chesney, pp. 524–48 (p. 527). Attempts to revive the breed began in the late 1800s, primarily using the Scottish Deerhound, which was believed to be descended from or even the same as the Irish wolfhound; G. A. Graham, *The Irish Wolfhound* (Dursley: Whitmore and Son, 1885), p. 1, and H. D. R., 'The Irish Wolf-Dog', *The Irish Penny Journal*, 1 (1841), 353–5 (p. 354).
106. Roderic O'Flaherty, *A Chorographical Description of West Or H-Iar Connaught, Written A.D. 1684*, ed. by James Hardiman (Dublin: Irish Archaeological Society, 1846), p. 180. Spelling modernised.
107. Ibid., p. 181.
108. Ibid., pp. 181–2.
109. C. H. Firth, 'Account of Money Spent in the Cromwellian Reconquest and Settlement of Ireland, 1649–1656', *The English Historical Review*, 14 (1899), 105–9 (p. 108).
110. Hickey, 'A Geographical Perspective on the Decline and Extermination of the Irish Wolf', pp. 190–1.
111. *Diary of Thomas Burton, Esq.*, ed. and trans. by John Towill Rutt, 4 vols (London: Colburn, 1828), II, pp. 210–11.
112. Hickey, 'A Geographical Perspective on the Decline and Extermination of the Irish Wolf', p. 195.
113. Hickey, *Wolves in Ireland*, pp. 103–4.
114. Fairley, *An Irish Beast Book*, p. 301.
115. Ibid., p. 302. The method was described as 'more than ordinary, and never known in this Kingdom' [spelling modernised].
116. Jesse, *Researches into the History of the British Dog*, II, p. 182.
117. Hickey, 'A Geographical Perspective on the Decline and Extermination of the Irish Wolf', p. 195.
118. Ibid., p. 197.
119. Hickey, *Wolves in Ireland*, p. 105.
120. Fairley, *An Irish Beast Book*, p. 303.
121. J. E. Harting, 'Wolves in Ireland', *Zoologist*, 3.9 (1885), p. 268.
122. Raye, *The Atlas of Early Modern Wildlife*, p. 45.
123. Hickey, 'A Geographical Perspective on the Decline and Extermination of the Irish Wolf', p. 197.
124. Fairley, *An Irish Beast Book*, pp. 306–8.

125. Ibid., p. 307, and Henry Morris et al., 'Miscellanea', *Béaloideas*, 10 (1940), 285–303 (pp. 287–8).
126. Fairley, *An Irish Beast Book*, p. 309.
127. Ibid, pp. 308–9.
128. Ibid., p. 305, and Hickey, *Wolves in Ireland*, p. 28.
129. Fairley, *An Irish Beast Book*, p. 309.
130. Gow, *Hunt for the Shadow Wolf*, p. 186. To view this painting, see Stephen Catterson Smith, 1806–1872, 'Portrait of John Henry Watson', Royal Dublin Society Library & Archives <https://digitalarchive.rds.ie/items/show/3987> [accessed 09 January 2025].
131. Fairley, *An Irish Beast Book*, p. 309; Nora Fisher, 'The Last Irish Wolf', *The Irish Naturalists' Journal*, 5 (1934), 41 (p. 41); and C. B. Moffat, 'The Mammals of Ireland', *Proceedings of the Royal Irish Academy Section B: Biological, Geological, and Chemical Science*, 44 (1937/1938), 61–128 (p. 75).
132. Fairley, *An Irish Beast Book*, pp. 309–10.
133. Barnett, *The Missing Lynx*, p. 243.
134. Fairley, *An Irish Beast Book*, pp. 309–10.
135. Anon., 'The British Wolf', *Spectator*, 26 January 1901, pp. 10–11 (p. 10), quoted in Fairley, *An Irish Beast Book*, p. 310.
136. Hickey, *Wolves in Ireland*, p. 47.
137. Harry Johnston, *British Mammals* (London: Hutchinson and Co., 1903), p. 128.
138. Gary Cunningham and Ronan Coghlan, *Mystery Animals of Ireland* (Woolsery: CFZ Press, 2000), pp. 147–52.
139. Keith Thomas, *Man and the Natural World: Changing Attitudes in England 1500–1800* (London: Allen Lane, 1983), p. 273.
140. Edward Topsell, *The Historie of Foure-Footed Beastes* (London: Iaggard, 1607), p. 626. Spelling modernised.
141. Laurent Brassart et al., 'Understanding and Controlling the Environment in Early Modern History (ca. 1500–1800)', in *The European Experience: A Multi-Perspective History of Modern Europe, 1500–2000*, ed. by Jan Hansen et al. (Cambridge: Open Book Publishers, 2023), pp. 529–38 (pp. 529–30 and 536–7).
142. *The Works of the Reverend Mr Edm. Hickeringill*, 2 vols (London: Bragge, 1709), I, p. 358 [spelling modernised], cited in Thomas, *Man and the Natural World*, p. 273.
143. Thomas, *Man and the Natural World*, p. 274.
144. Camden, 'Darbyshire', in *Britain, or a Chorographical Description of the Most Flourishing Kingdoms, England, Scotland, and Ireland*, trans. by Holland, pp. 553–9 (p. 556) [spelling modernised], cited in Dent, *Lost Beasts of Britain*, p. 13.
145. Philipp Camerarius, *The Living Librarie, or, Meditations and Observations Historical, Natural, Moral, Political, and Poetical*, trans. by John Molle (London: Islip, 1621), pp. 98–9 (p. 99), reproduced in Stillman, 'Philip Sidney, Thomas More, and Table Talk', pp. 349–50 (p. 349). Spelling modernised.
146. Stillman, 'Philip Sidney, Thomas More, and Table Talk', p. 346.
147. *Holinshed's Chronicles of England, Scotland and Ireland*, I: *England* (1807), p. 378.
148. Camerarius, *The Living Librarie*, p. 98, reproduced in Stillman, 'Philip Sidney, Thomas More, and Table Talk', p. 349. Spelling modernised.
149. *Holinshed's Chronicles of England, Scotland and Ireland*, I: *England* (1807), p. 378.
150. Caroline Grigson, *Menagerie: The History of Exotic Animals in England 1100–1837* (Oxford: Oxford University Press, 2016), p. 7.

151. Ibid., p. 50.
152. Midgley, 'The Problem of Living with Wildness', p. 185.
153. Grigson, *Menagerie*, p. 66.
154. Fychan, *Galwad y Blaidd*, pp. 126–7.
155. Grigson, *Menagerie*, p. 103. Spelling modernised.
156. Ibid., p. 84.
157. Ibid., pp. 157–8.
158. Hickey, *Wolves in Ireland*, pp. 62–3.
159. Ada Kathleen Longfield, *Anglo-Irish Trade in the Sixteenth Century* (London: Routledge & Sons, 1929), p. 64.
160. Faiers, *Fur: A Sensitive History*, p. 100.
161. Kathleen Walker-Meikle, 'Furs for Earls', *Renaissance Skin* <https://renaissanceskin.ac.uk/themes/consuming> [accessed 10 September 2023].
162. Camerarius, *The Living Librarie*, p. 99, reproduced in Stillman, 'Philip Sidney, Thomas More, and Table Talk', p. 350. Spelling modernised.
163. Camerarius, *The Living Librarie*, pp. 98 and 99, reproduced in Stillman, 'Philip Sidney, Thomas More, and Table Talk', p. 349. Spelling modernised.
164. Guy Miège, *The New State of England Under Their Majesties K. William and Q. Mary* (London: Robinson, 1691), p. 22.
165. Raye, *The Atlas of Early Modern Wildlife*, p. 44.
166. Philippe Glardon, 'The Relationship between Text and Illustration in Mid-Sixteenth-Century Natural History Treatises', trans. by Susan Becker, in *A Cultural History of Animals in the Renaissance*, ed. by Bruce Boehrer (London: Berg, 2007), pp. 119–45 (p. 135), translating from Pierre Belon, *La nature et diversité des poissons* (Paris: Estienne, 1555), p. 28.
167. Thomas Dekker, *Lantern and Candlelight (1608)*, ed. by Viviana Comensoli (Toronto: Centre for Reformation and Renaissance Studies, 2007), p. 77.
168. Thomas Wilson, *A Discourse Upon Usury by Way of Dialogue and Orations*, intro. by R. H. Tawney (London: Cass & Co., 1962), pp. 182–3. Spelling modernised.
169. William Shakespeare, *The Merchant of Venice*, ed. by John Drakakis (London: Arden Shakespeare, 2010), Act IV scene 1, lines. 132–7 (p. 343).
170. Alanna Skuse, *Constructions of Cancer in Early Modern England: Ravenous Natures* (Basingstoke: Palgrave Macmillan, 2015), p. 65.
171. Dana Rehn, 'The Wolf–Human Hybrid Motif Used in Protestant Propaganda Against the Re-Catholicism of England', *Dana K Rehn* (2022) <https://danakrehnblog.wordpress.com/2022/03/27/exiled-english-protestants-in-germany> [accessed 27 October 2025].
172. Skuse, *Constructions of Cancer in Early Modern England*, p. 65.
173. Topsell, *The Historie of Foure-Footed Beastes*, p. 746. Spelling modernised.
174. Skuse, *Constructions of Cancer in Early Modern England*, p. 63, citing Daniel Turner, *De Morbis Cutaneis: A Treatise of Diseases Incident to the Skin* (London: Bonwicke, Freeman, Goodwin, Walthoe, Wotton, Menship, Nicholson, Parker, Jooke, and Smith, 1714), p. 76.
175. Skuse, *Constructions of Cancer in Early Modern England*, pp. 65–6.
176. Williams, *Howls of Imagination*, p. 57, and Pluskowski, *Wolves and the Wilderness*, p. 174.
177. *A True Discourse. Declaring the Damnable Life and Death of One Stubbe Peeter*, trans. by George Bores (London: Venge, 1590) <http://name.umdl.umich.edu/A13085.0001.001> [accessed 27 October 2025]. Spelling modernised.
178. Ibid.
179. King James the First, *Daemonologie (1597)* (London: The Bodley Head, 1924), p. 61. Spelling modernised.

180. Robert Bayfield, 'A Treatise', in *A Lycanthropy Reader: Werewolves in Western Culture*, ed. by Charlotte F. Otten (Syracuse, NY: Syracuse University Press, 1986), p. 47.
181. Susan Wiseman, *Writing Metamorphosis in the English Renaissance: 1550–1700* (Cambridge: Cambridge University Press, 2014), pp. 156–8.
182. Topsell, *The Historie of Foure-Footed Beastes*, p. 745. Spelling modernised.
183. Ibid. Spelling modernised.
184. John Ogilby, *America: Being the Latest, and Most Accurate Description of the New World* (London: the author, 1671), pp. 8 and 15. Spelling modernised.
185. Marvin, *Wolf*, p. 93.
186. Alec Brownlow, 'A Wolf in the Garden: Ideology and Change in the Adirondack Landscape', in *Animal Spaces, Beastly Places: New Geographies of Human–Animal Relations*, ed. by Chris Philo and Chris Wilbert (London and New York: Routledge, 2000), pp. 143–60 (pp. 148–9 and 150).
187. Valerie M. Fogleman, 'American Attitudes Towards Wolves: A History of Misperception', *Environmental Review*, 13.1 (1989), 63–94 (p. 64).
188. Marvin, *Wolf*, p. 89.
189. Brett L. Walker, 'Animals and the Intimacy of History', *History and Theory*, 52.4 (2013), 45–67 (p. 66).
190. Marvin, *Wolf*, p. 101.
191. 'Ontario County: Reminiscences of James Sperry', in *The Development of Central and Western New York: From the Arrival of the White Man to the Eve of the Civil War as Portrayed in Contemporary Accounts*, ed. by Clayton Mau, rev. edn (Dansville, NT: Owen, 1958), pp. 121–4 (p. 121).
192. William Wood, *New Englands Prospect: A True, Lively, and Experimentall Description of That Part of America, Commonly Called New England* (London: Cotes, 1634), pp. 44 and 24. Spelling modernised.
193. Fogleman, 'American Attitudes Towards Wolves', pp. 64–5.
194. Marvin, *Wolf*, p. 101.
195. Fogleman, 'American Attitudes Towards Wolves', p. 65.
196. Wood, *New Englands Prospect*, p. 21. Spelling modernised.
197. Marvin, *Wolf*, p. 87.
198. Letter from Edmund Browne to Sir Simonds D'Ewes (September 7, 1638), in *Letters from New England: The Massachusetts Bay Colony, 1629–1638*, ed. by Everett Emerson (Amherst: University of Massachusetts Press, 1976), pp. 224–30 (p. 228).
199. Wood, *New Englands Prospect*, p. 23. Spelling modernised.
200. Jon T. Coleman, *Vicious: Wolves and Men in America* (New Haven, CT: Yale University Press, 2004), p. 37.
201. Fogleman, 'American Attitudes Towards Wolves', p. 66.
202. Ibid., pp. 66 and 67
203. Wood, *New Englands Prospect*, p. 24.
204. Fogleman, 'American Attitudes Towards Wolves', pp. 66–7 and 86 n. 13.
205. Ibid., p. 67.
206. Marvin, *Wolf*, pp. 84–5.
207. Fogleman, 'American Attitudes Towards Wolves', pp. 66 and 67.
208. Ibid., p. 67.
209. Marvin, *Wolf*, p. 86.
210. Peter Arnds, *Wolves at the Door: Migration, Dehumanization, Rewilding the World* (New York and London: Bloomsbury, 2021), p. 56.
211. Brownlow, 'A Wolf in the Garden', pp. 150–1.
212. Ibid., p. 151.

213. Ibid., pp. 150–1.
214. Thomas, *Man and the Natural World*, p. 47.
215. *The Statutes at Large; Being a Collection of All the Laws of Virginia, From the First Session of the Legislature, in the Year 1619, Volume 1*, ed. by William Waller Hening (Richmond, VA: Pleasants, 1809), p. 395.
216. Arnds, *Wolves at the Door*, pp. 54–5.
217. Theodore Roosevelt, *Hunting the Grisly and Other Sketches* (New York: Review of Reviews Company, 1904), p. 213.
218. Ibid., p. 232.
219. Marvin, *Wolf*, p. 107.
220. Ibid., pp. 108–9.

Chapter 6: A Post-wolf World

1. Smout, MacDonald, and Watson, *A History of the Native Woodlands of Scotland*, pp. 122–3.
2. Crumley, *The Last Wolf*, p. 56.
3. Benedict MacDonald, *Rebirding: Rewilding Britain and Its Birds* (Exeter: Pelagic Publishing, 2019), pp. 35–6, and Eric Richards, *The Highland Clearances: People, Landlords and Rural Turmoil* (Edinburgh: Birlinn, 2002), p. 289.
4. MacDonald, *Rebirding*, pp. 40–1.
5. Richards, *The Highland Clearances*, p. 59.
6. Ernest Barker, *Ireland in the Last Fifty Years (1866–1918)*, second edn (Oxford: Clarendon Press, 1919), pp. 44–5.
7. Thomas More, *Utopia*, trans. by Clarence H. Miller, second edn (New Haven, CT: Yale University Press, 2014), pp. 22–3.
8. Bruce Boehrer, *Environmental Degradation in Jacobean Drama* (Cambridge: Cambridge University Press, 2013), p. 9, and J. R. Wordie, 'The Chronology of English Enclosure, 1500–1914', *The Economic History Review*, 36 (1983), 483–505 (p. 494).
9. Boehrer, *Environmental Degradation in Jacobean Drama*, p. 9.
10. Michael Turner, 'Counting Sheep: Waking Up to New Estimates of Livestock Numbers in England *c.*1800', *Agricultural History Review*, 46 (1998), 142–61 (p. 159, Table 6).
11. Winder, *The Last Wolf*, pp. 256–7.
12. Ibid., pp. 250–2.
13. *The Landscape Gardening and Landscape Architecture of the Late Humphry Repton*, ed. by J. C. Loudon, new edn (London: Longman and Co., 1840), p. 531.
14. Stanley Baldwin, *On England, and Other Addresses* (London: Allan & Co., 1926), p. 7, quoted in Winder, *The Last Wolf*, p. 406.
15. 'One in Eight Young People Have Never Seen a Cow in Real Life', *Telegraph*, 31 July 2017 <https://www.telegraph.co.uk/news/2017/07/31/one-eight-young-people-have-never-seen-cow-real-life> [accessed 20 January 2025].
16. Colin Matheson, 'The Grey Wolf', *Antiquity*, 17 (1943), 11–18 (p. 17).
17. Terry Hooper-Scharff, *The Red Paper: Canids* (Bristol: Black Tower Books, 2011), p. 127.
18. 'Editorial', *Western Times*, 28 November 1887, p. 4, cited in Helen Cowie, 'Wolves Behind Bars', in *The Wolf: Culture, Nature, Heritage*, ed. by Convery et al., pp. 57–67 (p. 59).
19. Cowie, 'Wolves Behind Bars', p. 60.
20. Ibid., pp. 58–9, citing 'The Zoo in Mourning: Sally's Death', *Newcastle Weekly Courant*, 12 September 1891, p. 5.

21. 'The Extraordinary Escape of a Wolf from the Tower of London', *Standard*, 29 April 1834, p. 1, quoted in Cowie, 'Wolves Behind Bars', pp. 60–1
22. 'Escape of Wolves at Sanger's', *Standard*, 13 February 1888, p. 3, quoted in Cowie, 'Wolves Behind Bars', p. 61.
23. Williams, *Howls of Imagination*, p. 10.
24. 'Extraordinary Scene at the Zoological Gardens, Dublin', *Liverpool Mercury*, 13 January 1858, p. 2, cited in Cowie, 'Wolves Behind Bars', p. 57.
25. 'Escape of a Wolf', *Bristol Mercury*, 1 July 1843, p. 2, cited in Cowie, 'Wolves Behind Bars', p. 60.
26. Hooper-Scharff, *The Red Paper: Canids*, p. 129.
27. 'A Wolf Shot at Peckham', *Morning Advertiser*, 2 January 1847, p. 4, quoted in Hooper-Scharff, *The Red Paper: Canids*, p. 130 (though erroneously attributed to the *Bristol Mercury*).
28. Charles Fort, *Lo!*, in *The Complete Books of Charles Fort* (New York: Dover Publications, 1974), pp. 539–839 (pp. 649–61), and 'The Marauding Wolf in Northumberland', *Times*, 21 December 1904, p. 7.
29. The Dangerous Wild Animals Act, which stipulated that UK residents could no longer keep wolves and other wild animals as pets without a licence, was passed in 1976.
30. Felix Allen, 'WHERE WOLF? Mystery as Dad Records 'WOLVES' Howling in the Night 'Like a Horror Film' Near Scottish Hotel', *Sun*, 14 September 2017 <https://www.thesun.co.uk/news/4463828/hotel-manager-scotland-wolf-outer-hebrides-video> [accessed 02 November 2023].
31. Matthew Weaver, 'Wolf Experts Urge UK Police Not to Shoot Escaped Animal', *Guardian*, 18 January 2018 <https://www.theguardian.com/uk-news/2018/jan/18/wolf-escapes-wildlife-sanctuary-near-school-berkshire> [accessed 09 October 2023].
32. Barnett, *The Missing Lynx*, p. 253.
33. Patrick Barkham, 'Who's Afraid of the Big Bad Escaped Wolf? Not Me', *Guardian*, 19 January 2018 <https://www.theguardian.com/commentisfree/2018/jan/19/afraid-big-bad-wolf-torak-humans> [accessed 06 November 2023].
34. The year prior, a female wolf who escaped from Cotswold Wildlife Park shortly after giving birth to a litter of pups was shot dead.
35. Jay M. Smith, *Monsters of the Gévaudan: The Making of a Beast* (Cambridge, MA: Harvard University Press, 2011), pp. 39 and 41.
36. Ibid., p. 41.
37. Ibid., p. 207.
38. Ibid., pp. 208–9.
39. Ibid., p. 226.
40. Ibid., p. 240.
41. Ibid., p. 241.
42. Linnell et al., 'The Fear of Wolves', p. 19.
43. Dent, *Lost Beasts of Britain*, p. 121.
44. Hooper-Scharff, *The Red Paper: Canids*, p. 43.
45. 'Fox Hunting', *Morning Post*, 14 February 1810, p. 3, quoted in Hooper-Scharff, *The Red Paper: Canids*, p. 165.
46. s.v. 'rabid', *Oxford English Dictionary Online* (2024) <https://www.oed.com/dictionary/rabid_adj?tab=meaning_and_use> [accessed 07 January 2025].
47. Crumley, *The Last Wolf*, p. 50.
48. Ibid. In the late eighteenth century, a budding Welsh entrepreneur named David Pritchard apparently capitalised upon this tendency to great profit. The new landlord of a hotel in the Gwynedd village of Beddgelert, Pritchard hatched a plan to draw visitors to the village and, by extension, customers to

his hotel. And nothing was more compelling than a story of ravening wolves and regret.

Pritchard invented the character of Gelert, a loyal hound belonging to Prince Llewellyn of Wales (r.1199–1240), who was tragically killed by his master when he was discovered, covered in blood, next to the empty cradle of Llewellyn's infant son. Only after he had dispatched his dog did Llewellyn find his baby, unharmed, beside the body of a wolf that Gelert had killed to protect his infant charge. Llewellyn buried his poor, brave hound in the village, whose grave can still be seen there today, and who was memorialised in the name of Beddgelert.

Or so the story goes. The name Beddgelert actually comes from a saint by the name of Celert, and Pritchard himself placed the grave in the village, a ready-made tourist destination to drum up business. It worked wonders, especially when the story of the brave dog was immortalised by the English poet William Spencer (1769–1834).

While it is possible that the story may have existed prior to Pritchard's enterprising efforts, its fantastical nature is not in question. The tale of a loyal animal defending its infant charge against a nefarious intruder is ancient, found in various versions from all over the world. One of the earliest is from India, and features a pet mongoose who kills a snake that attacks the baby it is guarding. The story travelled westwards and, at some point, the main character became a dog – a very similar story to Gelert's, from France, tells the tale of a greyhound named Guinefort, who killed a snake that attacked his owner's child.

It is telling of the readiness with which wolves are mythologised in Britain that the assailant became a wolf in the Welsh version.

49. Research into news reports of two canid attacks on people found that the media covered these incidents more fervently (and less accurately) when wolves were considered to be the culprits. They were much less interested in both cases when it transpired that dogs were the attackers; Ugo Arbieu et al., 'News Selection and Framing: The Media as a Stakeholder in Human–carnivore Coexistence', *Environmental Research Letters*, 16.6 (2021).
50. Christopher Snowdon, 'Reintroducing Wolves to Britain is Pure Insanity', *The Spectator*, 15 August 2023 <https://www.spectator.co.uk/article/bringing-back-wolves-isnt-the-answer-to-rewilding-britain/> [accessed 29 October 2023].
51. Richard Hartley-Parkinson, 'Girl, 12, Kidnapped and Gang Raped by Men Labelled "Ravenous Pack of Wolves"', *Metro*, 13 April 2018 <https://metro.co.uk/2018/04/13/girl-12-kidnapped-gang-raped-men-labelled-ravenous-pack-wolves-7463382/> [accessed 25 October 2023], and Samuel Jones, '"Wild Bunch of Wolves": Anger and Trepidation at Heart of Acklam's Yob-plagued Street', *Teesside Live*, 21 March 2021 <https://www.gazettelive.co.uk/news/teesside-news/wild-bunch-wolves-anger-trepidation-20197989> [accessed 25 October 2023].
52. Micah Halpern, 'We Must Track and Trap Lone Wolf Terrorists', *The Observer*, 25 November 2014 <https://observer.com/2014/11/we-must-track-and-trap-lone-wolf-terrorists/> [accessed 27 October 2023].
53. Even the climate emergency was described as the 'big and bad' 'weather wolf' by then-Chief Executive of the Environment Agency, Sir James Bevan; 'Watching the Wolf: Why the Climate Emergency Threatens Us All', Environment Agency blog, 25 February 2021 <https://environmentagency.blog.gov.uk/2021/02/25/watching-the-wolf-why-the-climate-emergency-threatens-us-all/> [accessed 05 March 2021].
54. Lopez, *Of Wolves and Men*, p. 140.
55. Arnds, *Wolves at the Door*, pp. 1–2.

56. A graphic circulated on social media accounts belonging to American pro-hunting group Hunter Nation described how '216 Wisconsin wolves' 'killed 4,320 deer', which equalled '650,000 lbs of meat' and '2,592,000 meals' 'taken from Wisconsin families'.
57. Arnds, *Wolves at the Door*, pp. 2 and 143.
58. Ibid., pp. 3–5 and 174.
59. L. David Mech, Morgan Anderson, and H. Dean Cluff, *The Ellesmere Wolves: Behavior and Ecology in the High Arctic* (Chicago: University of Chicago Press, 2025), p. 214.
60. Adam Lusher, 'Donald Trump Supporters Tell Immigrants "The Wolves are Coming, You Are the Hunted" as Race Hate Fears Rise', *Independent*, 9 November 2016 <https://www.the-independent.com/news/world/americas/us-politics/donald-trump-wins-racist-racism-race-hate-immigrants-nigel-farage-ukip-brexit-post-referendum-illegals-mexicans-build-wall-the-wolves-are-coming-you-are-the-hunted-kkk-white-power-latinos-blacks-a7407951.html> [accessed 26 October 2023].
61. Arnds, *Wolves at the Door*, p. 3.
62. Marvin, *Wolf*, p. 76.
63. 'Racist "Wolves" Emerge from Shadows', *BBC News*, 25 April 1999 <http://news.bbc.co.uk/1/hi/uk/328181.stm> [accessed 24 October 2023].
64. Charles Perrault, 'Little Red Riding-Hood', in *The Complete Fairy Tales*, trans. by Christopher Betts (Oxford: Oxford University Press, 2009), pp. 99–103.
65. 'Little Red Cap', in Brothers Grimm, *The Complete Fairy Tales*, ed. and trans. by Jack Zipes (London: Vintage, 2007), pp. 125–9.
66. Marvin, *Wolf*, p. 68.
67. Lopez, *Of Wolves and Men*, p. 140.
68. Jeanne Dubino, 'Mad Dogs and Irishmen: Dogs, *Dracula*, and the Colonial Irish Other', in *Animals in Irish Literature and Culture*, ed. by Kathryn Kirkpatrick and Borbála Faragó (Basingstoke: Palgrave Macmillan, 2015), pp. 199–213 (p. 204).
69. Arnds, *Wolves at the Door*, p. 77.
70. Pavol Prokop, Muhammet Usak, and Mehmet Erdogan, 'Good Predators in Bad Stories: Cross-Cultural Comparison of Children's Attitudes Towards Wolves', *Journal of Baltic Science Education*, 10 (2011), 229–42 (p. 237).
71. Ibid., pp. 238–9.
72. Ellis's work has been denounced by wolf biologists and conservation organisations. One member of the UK Wolf Conversation Trust described his approach as 'going down the same route as the macho-driven pursuits of the seemingly burgeoning number of TV presenters who think it is acceptable to drag what are perceived to be scary and highly dangerous creatures from their habitats, and wrestle with them and dominate them'; Denise Taylor, 'Humans Are Not Wolves!', *Wolf Print*, 31 (2007), pp. 14–15 (p. 14).
73. Bram Stoker, *Dracula* (Oxford: Oxford University Press, 2011), pp. 16 and 21.
74. Chad Richardson, 'Depiction of Wolves in New Fortnite Release is Inaccurate, Irresponsible and Could Have Lasting Consequences', International Wolf Center <https://wolf.org/media-releases/depiction-of-wolves-in-new-fortnite-release-is-inaccurate-irresponsible-and-could-have-lasting-consequences> [accessed 14 November 2023].
75. Ibid.
76. Marvin, *Wolf*, p. 137.
77. John Lockwood Kipling, *Beast and Man in India: A Popular Sketch of Indian Animals in Their Relations with the People* (London: Macmillan and Co., 1892), p. 281.
78. For more on sympathetic wolf stories, see Peter Hollindale, 'Why the Wolves Are Running', *The Lion and the Unicorn*, 23 (1999), 97–115.

79. Lucy Jones, *Foxes Unearthed: A Story of Love and Loathing in Modern Britain* (London: Elliott and Thompson, 2016), p. 30.
80. Ibid., p. 11.
81. Erin Cunningham, 'Wild Alphabet: The Wolf in Irish Poetry', *The Learned Pig* (2017) <https://thelearnedpig.org/wild-alphabet-wolf-irish-poetry> [accessed 20 August 2023].
82. In his poem 'Midnight', Seamus Heaney contends with these losses in parallel, describing how Irish wolf-land has been tamed. The forests are 'coopered to wine casks', the 'old dens' of the wolves 'are soaking', flooded, and the 'panting, lolling, / Vapouring' animals have been replaced by 'small vermin' which scuttle about (perhaps a metaphor for the English themselves). The result of this natural and cultural taming is the silencing of Irish tongues, with Heaney's itself 'Leashed in [his] throat'. In losing the wolf, Heaney suggests, the Irish lost their wildness, freedom, and perhaps even themselves; Seamus Heaney, 'Midnight', in *Wintering Out* (London: Faber & Faber, 1972), p. 35. See Potts, *Contemporary Irish Writing and Environmentalism*, pp. 146–7.
83. Clarissa Pinkola Estés, *Women Who Run with the Wolves: Contacting the Power of the Wild Woman* (London: Rider, 2008), p. 2.
84. Ibid.
85. Marvin, *Wolf*, pp. 153–4.
86. Ibid., p. 115, quoting George M. Wright and Ben H. Thompson, *Fauna of the National Parks of the United States*, Fauna Series No. 2 – July 1934 (Washington, DC: United States Government Printing Office, 1935), p. 15.
87. Karen Jones, 'From Big Bad Wolf to Ecological Hero: *Canis Lupus* and the Culture(s) of Nature in the American–Canadian West', *American Review of Canadian Studies*, 40 (2010), 338–50 (p. 340).
88. Fogleman, 'American Attitudes Towards Wolves', p. 69.
89. Dan Flores, *Coyote America: A Natural and Supernatural History* (New York: Basic Books, 2016), p. 97.
90. E. A. Goldman, 'The Predatory Mammal Problem and the Balance of Nature', *Journal of Mammalogy*, 6 (1925), 28–33 (p. 33), and Stanley P. Young and Edward A. Goldman, *The Wolves of North America, Part 1: Their History, Life Habits, Economic Status, and Control* (New York: Dover, 1944), p. 1.
91. United States Department of Agriculture, Forest Service, 'Report of the Forester' (1927), p. 31.
92. Jones, 'From Big Bad Wolf to Ecological Hero', p. 343.
93. Matthew Wynn Sivils, 'Introduction', in Paul L. Errington, *Of Wilderness and Wolves*, ed. by Matthew Wynn Sivils (Iowa City: University of Iowa Press, 2015), pp. 1–24 (p. 4).
94. Jones, 'From Big Bad Wolf to Ecological Hero', p. 343.
95. Richard West Sellars, *Preserving Nature in the National Parks: A History* (New Haven, CT: Yale University Press, 2009), pp. 122–3.
96. Karen Jones, 'Writing the Wolf: Canine Tales and North American Environmental-Literary Tradition', *Environment and History*, 17 (2011), 201–28 (p. 213).
97. Sellars, *Preserving Nature in the National Parks*, p. 123.
98. Aldo Leopold, *A Sand County Almanac and Sketches Here and There* (London: Oxford University Press, 1970), pp. 130–2.
99. Fogleman, 'American Attitudes Towards Wolves', p. 69.
100. Greg Quill, 'Farley Mowat's Legacy: Our Supreme Storyteller', *Toronto Star*, 11 May 2012 <https://www.thestar.com/entertainment/books/farley-mowat-s-legacy-our-supreme-storyteller/article_ebbd7469-d78a-570d-aa4b-515556b6bba5.html> [accessed 15 November 2023].

101. Farley Mowat, *Never Cry Wolf* (New York: Back Bay Books, 2001), pp. 14–16.
102. Ibid., p. 76.
103. Stephanie Rutherford, *Villain, Vermin, Icon, Kin: Wolves and the Making of Canada* (Montreal & Kingston: McGill-Queen's University Press, 2022), p. 61.
104. Karen Jones, '*Never Cry Wolf*: Science, Sentiment, and the Literary Rehabilitation of *Canis Lupus*', *Canadian Historical Review*, 84 (2003), 1–16 (p. 2).
105. Ibid., p. 7.
106. Ibid., p. 15, and Mowat, *Never Cry Wolf*, pp. 102–7.
107. Linnell and Alleau, 'Predators That Kill Humans', p. 360.
108. In such packs, the breeding male and female are naturally the group's 'leaders', rather than individuals who have fought to attain status as 'top dog'. Older offspring from previous litters, being more mature and helping to care for new pups, are also naturally more 'dominant' than their siblings from younger litters. For pups in the same litter, conflict is not related to hierarchy but merely a matter of sibling squabbles. In these family-based groups, food is not fought over tooth and claw to reinforce hierarchies, but shared; see L. David Mech, 'Alpha Status, Dominance, and Division of Labor in Wolf Packs', *Canadian Journal of Zoology*, 77 (1999), 1196–1203, and Jane M. Packard, 'Wolf Behavior: Reproductive, Social, and Intelligent', in *Wolves: Behavior, Ecology, and Conservation*, ed. by Mech and Boitani, pp. 35–65 (pp. 54–5).
109. Marvin, *Wolf*, pp. 31–3; Mech and Boitani, 'Wolf Social Ecology', pp. 1–3; Packard, 'Wolf Behavior: Reproductive, Social, and Intelligent', p. 53; and Rivka Galchen, 'The Myth of the Alpha Wolf', *New Yorker*, 25 March 2023 <https://www.newyorker.com/science/elements/the-myth-of-the-alpha-wolf> [accessed 01 May 2024].
110. L. David Mech, 'Is Science in Danger of Sanctifying the Wolf?', *Biological Conservation*, 150 (2012), 143–9 (p. 146).
111. N. Thompson Hobbs et al., 'Does Restoring Apex Predators to Food Webs Restore Ecosystems? Large Carnivores in Yellowstone as a Model System', *Ecological Monographs*, 94 (2024), 1–32.
112. Whereby the return (or removal) of an animal at the top of the food chain produces a trickle-down effect which affects the entire ecosystem: wolves prey on deer, relieving pressure on the deer's food source, which in turn affects other species which live in or depend upon that food source.
113. Mech, 'Is Science in Danger of Sanctifying the Wolf?', p. 147.
114. Ibid., p. 146.
115. Ibid.
116. Ibid., p. 144, citing Justina C. Ray et al., 'Is Large Carnivore Conservation Equivalent to Biodiversity Conservation and How Can We Achieve Both?', in *Large Carnivores and the Conservation of Biodiversity*, ed. by Justina C. Ray et al. (Washington, DC: Island Press, 2005), pp. 400–27 (p. 426).
117. Arthur Middleton, 'Is the Wolf a Real American Hero?', *New York Times*, 9 March 2014 <https://www.nytimes.com/2014/03/10/opinion/is-the-wolf-a-real-american-hero.html> [accessed 18 March 2024]; Rene Beyers et al., 'The Wolves of Yellowstone: Saviours of the Songbird or Pieces of the Puzzle?', in *The Wolf: Culture, Nature, Heritage*, ed. by Convery et al., pp. 249–58 (pp. 252–3); and Virginia Morell, 'Predation, Not Fear of Wolves, Keeps Elk from Denuding Yellowstone', *Science*, 23 October 2024 <https://www.science.org/content/article/predation-not-fear-wolves-keeps-elk-denuding-yellowstone> [accessed 25 October 2024].
118. John Terborgh et al., 'The Role of Top Carnivores in Regulating Terrestrial Ecosystems', in *Continental Conservation: Scientific Foundations of Regional Reserve*

Networks, ed. by Michael E. Soulé and John Terborgh (Washington, DC: Island Press, 1999), pp. 39–64 (p. 58).

119. J. M. Alston et al., 'Reciprocity in Restoration Ecology: When Might Large Carnivore Reintroduction Restore Ecosystems?', *Biological Conservation*, 234 (2019), 82–9.
120. In parts of the United States where wolves have returned, the narrative of the 'saviour' wolf has created ill-feeling among some people, who in turn have 'popularize[d] their own myths about the reintroduced wolves', framing them as 'a voracious, nonnative strain' that 'devastate elk herds, spread elk diseases, and harass elk relentlessly—often just for fun', and about which the government tells packs of lies; Middleton, 'Is the Wolf a Real American Hero?'.
121. Mech, 'Is Science in Danger of Sanctifying the Wolf?', p. 147.
122. Ibid.

Chapter 7: Rewolfing

1. Their lack of natural predators is not the only cause of the burgeoning deer populations, though it is an important one. Other factors include greater forest cover thanks to land abandonment and planting efforts, a warming climate, and reduced numbers of sheep with which to compete. There are also increased controls and regulations surrounding their hunting, and their numbers are generally underestimated, meaning fewer deer are shot than is necessary to curb their increase. There is also little incentive for landowners who make money from deer stalking to cull them; R. J. Fuller and R. M. A. Gill, 'Ecological Impacts of Increasing Numbers of Deer in British Woodland', *Forestry*, 74 (2001), 193–9 (pp. 194–5).
2. MacDonald, *Rebirding*, p. 146.
3. The current populations of fallow deer in Britain and Ireland were introduced by the Normans, but the species was found in Britain during the Pleistocene and was also briefly present during Roman occupation. In Ireland, non-native roe deer were introduced from Scotland in the nineteenth century, but did not become established. In recent years, however, there have been a handful of sightings in eastern Ireland. Chinese water deer are also not found in Ireland, but a hybrid species of red and sika deer has been recorded in both Ireland and Britain.
4. Parliamentary Office of Science and Technology, 'Wild Deer', POSTnote 325 (2009) <https://post.parliament.uk/research-briefings/post-pn-325/> [accessed 20 April 2024].
5. See Robin Gill, 'The Impact of Deer on Woodland Biodiversity', Forestry Commission Information Note 36 (2000) <https://cdn.forestresearch.gov.uk/2000/01/fcin036.pdf> [accessed 04 July 2024], and Paul Dolman et al., 'Escalating Ecological Impacts of Deer in Lowland Woodland', *British Wildlife*, 21.4 (2010), 242–54.
6. Fuller and Gill, 'Ecological Impacts of Increasing Numbers of Deer', p. 196.
7. Gill, 'The Impact of Deer on Woodland Biodiversity', pp. 2–3.
8. Vashti Gwynn and Elias Symeonakis, 'Rule-based Habitat Suitability Modelling for the Reintroduction of the Grey Wolf (*Canis lupus*) in Scotland', *PLoS ONE*, 17 (2022), 1–21 (p. 2), and Gill, 'The Impact of Deer on Woodland Biodiversity', pp. 3–4. Current deer densities in Ireland are unknown, although recent years have seen record numbers of deer being culled, and it is thought that population and range are both growing rapidly. Populations in Wales and England are also thought to be expanding in both numbers and range.

9. Reforesting Scotland, 'The Impact and Management of Deer in Scotland' <https://reforestingscotland.org/influencing-policy/the-impact-and-management-of-deer-in-scotland> [accessed 03 July 2024]; Tom Edwards and Wendy Kenyon, 'Wild Deer in Scotland', Scottish Parliament Information Centre (SPICe) Briefing 13/74 (8 November 2013), p. 19; and Christopher Sandom et al., 'Exploring the Value of Wolves (*Canis lupus*) in Landscape-Scale Fenced Reserves for Ecological Restoration in the Scottish Highlands', in *Fencing for Conservation: Restriction of Evolutionary Potential or a Riposte to Threatening Processes?*, ed. by Michael J. Somers and Matthew Hayward (New York: Springer, 2012), pp. 245–76 (p. 248).
10. MacDonald, *Rebirding*, p. 138.
11. Andrés Ordiz, Richard Bischof, and Jon E. Swenson, 'Saving Large Carnivores, but Losing the Apex Predator?', *Biological Conservation*, 168 (2013), 128–33 (p. 131), and Kristy M. Ferraro and Christopher Hirst, 'Missing Carcasses, Lost Nutrients: Quantifying Nutrient Losses from Deer Culling Practices in Scotland', *Ecological Solutions and Evidence*, 5 (2024).
12. Studies of wolves in Europe suggest that they prefer red deer over other ungulates when this species is available. There are certainly plentiful red deer in the Scottish Highlands, where wolves are most likely to be reintroduced; Gwynn and Symeonakis, 'Rule-based Habitat Suitability Modelling', p. 7, and Sandom et al., 'Exploring the Value of Wolves in Landscape-Scale Fenced Reserves', p. 268.
13. D. W. Yalden, 'The Problems of Reintroducing Carnivores', *Symposia of the Zoological Society of London*, 65 (1993), 289–306 (p. 293).
14. Charles J. Wilson, 'Could We Live with Reintroduced Large Carnivores in the UK?', *Mammal Review*, 34 (2004), 211–32 (p. 218).
15. Ibid., p. 219, citing Juan Carlos Blanco, Santiago Reig, and Luis de la Cuesta, 'Distribution, Status and Conservation Problems of the Wolf in Spain', *Biological Conservation*, 60 (1992), 73–80.
16. L. David Mech, 'A New Era for Carnivore Conservation', *Wildlife Society Bulletin*, 24 (1996), 397–401 (p. 399).
17. Adrian D. Manning, Iain J. Gordon, and William J. Ripple, 'Restoring Landscapes of Fear with Wolves in the Scottish Highlands', *Biological Conservation*, 142 (2009), pp. 2314–21 (p. 2316).
18. John W. Laundré, Lucina Hernández, and Kelly B. Altendorf, 'Wolves, Elk, and Bison: Reestablishing the "Landscape of Fear" in Yellowstone National Park, U.S.A.', *Canadian Journal of Zoology*, 79 (2001), 1401–9 (p. 1409), and Marek C. Allen, Michael Clinchya, and Liana Y. Zanette, 'Fear of Predators in Free-living Wildlife Reduces Population Growth Over Generations', *Proceedings of the National Academy of Sciences*, 119 (2022), 1–6.
19. Liana Y. Zanette and Michael Clinchy, 'Ecology and Neurobiology of Fear in Free-Living Wildlife', *Annual Review of Ecology, Evolution, and Systematics*, 51 (2020), 297–318.
20. D. P. J. Kuijper et al., 'Landscape of Fear in Europe: Wolves Affect Spatial Patterns of Ungulate Browsing in Białowieża Primeval Forest, Poland', *Ecography*, 36 (2013), 1263–75 (p. 1270).
21. See, for example, Robert L. Beschta and William J. Ripple, 'Can Large Carnivores Change Streams Via a Trophic Cascade?', *Ecohydrology*, 12 (2018), 1–13; Robert L. Beschta and William J. Ripple, 'Riparian Vegetation Recovery in Yellowstone: The First Two Decades After Wolf Reintroduction', *Biological Conservation*, 198 (2016), 93–103; William J. Ripple and Robert L. Beschta, 'Wolf Reintroduction, Predation Risk, and Cottonwood Recovery in Yellowstone National Park', *Forest Ecology and Management*, 184 (2003), 299–313; William

J. Ripple and Robert L. Beschta, 'Wolves and the Ecology of Fear: Can Predation Risk Structure Ecosystems?', *BioScience*, 54 (2004), 755–66; and William J. Ripple and Robert L. Beschta, 'Trophic Cascades in Yellowstone: The First 15 Years After Wolf Reintroduction', *Biological Conservation*, 145 (2012), 205–13. See also pp. 179–80 above.

22. Arthur D. Middleton et al., 'Linking Anti-predator Behaviour to Prey Demography Reveals Limited Risk Effects of an Actively Hunting Large Carnivore', *Ecology Letters*, 16 (2013), 1023–30.
23. Michel T. Kohl et al., 'Diel Predator Activity Drives a Dynamic Landscape of Fear', *Ecological Monographs* 88 (2018), 638–52; N. Thompson Hobbs et al., 'Does Restoring Apex Predators to Food Webs Restore Ecosystems? Large Carnivores in Yellowstone as a Model System', *Ecological Monographs*, 94 (2024); and Middleton, 'Is the Wolf a Real American Hero?'.
24. Middleton et al., 'Linking Anti-predator Behaviour to Prey Demography'.
25. Kuijper et al., 'Landscape of Fear in Europe', p. 1263.
26. Hugh Webster, 'What if Wolves Don't Change Rivers, or the Lynx Lacks Bite? Rethinking a Rewilding Orthodoxy', *British Wildlife*, 33.2 (2021), pp. 91–7 (pp. 94–5).
27. Giorgia Ausilio et al., 'Ecological Effects of Wolves in Anthropogenic Landscapes: The Potential for Trophic Cascades is Context-Dependent', *Frontiers in Ecology and Evolution*, 8 (2021).
28. One study suggested that defensive prey behaviour took only one generation to be reinstated after the return of predators; Joel Berger, 'Carnivore Repatriation and Holarctic Prey: Narrowing the Deficit in Ecological Effectiveness', *Conservation Biology*, 21 (2007), 1105–16.
29. Erlend B. Nilsen et al., 'Wolf Reintroduction to Scotland: Public Attitudes and Consequences for Red Deer Management', *Proceedings of the Royal Society B*, 274 (2007), 995–1002.
30. D. V. Spracklen et al., 'Wolf Reintroduction to Scotland Could Support Substantial Native Woodland Expansion and Associated Carbon Sequestration', *Ecological Solutions and Evidence*, 6 (2025).
31. Joseph W. Bull et al., 'Fences Can Support Restoration in Human-dominated Ecosystems When Rewilding with Large Predators', *Restoration Ecology*, 27 (2019), 198–209.
32. Martyn L. Gorman, 'Restoring Ecological Balance to the British Mammal Fauna', *Mammal Review*, 37 (2007), 316–25 (p. 322).
33. Ibid., pp. 322–4.
34. Manning, Gordon and Ripple, 'Restoring Landscapes of Fear with Wolves in the Scottish Highlands'.
35. Bull et al., 'Fences Can Support Restoration in Human-dominated Ecosystems', pp. 9–10.
36. Smith and Ferguson, *Decade of the Wolf*, p. 130, and Christopher C. Wilmers et al., 'Trophic Facilitation by Introduced Top Predators: Grey Wolf Subsidies to Scavengers in Yellowstone National Park', *Journal of Animal Ecology*, 72 (2003), 909–16 (p. 914).
37. Arnds, *Wolves at the Door*, p. 175.
38. Staffan Roos et al., 'A Review of Predation as a Limiting Factor for Bird Populations in Mesopredator-rich Landscapes: A Case Study of the UK', *Biological Reviews*, 93 (2018), 1915–37 (p. 1920).
39. J. M. Alston et al., 'Reciprocity in Restoration Ecology: When Might Large Carnivore Reintroduction Restore Ecosystems?', *Biological Conservation*, 234 (2019), 82–9.
40. Mech, 'A New Era for Carnivore Conservation', p. 398.

41. Terborgh et al., 'The Role of Top Carnivores in Regulating Terrestrial Ecosystems', p. 58.
42. Chris Sandom et al., 'Rewilding', in *Key Topics in Conservation Biology 2*, ed. by David W. Macdonald and Katherine J. Willis (Chichester: Wiley & Sons, 2013), pp. 430–51 (p. 434).
43. MacDonald, *Rebirding*, p. 163.
44. Dana Hoag et al., 'Economic Consequences of the Wolf Comeback in the Western United States', *Western Economics Forum*, 20.1 (2022), 61–70 (p. 65).
45. Colwell, *Beak, Tooth and Claw*, pp. 191–2.
46. *Wolves: Proceedings of the First Working Meeting of Wolf Specialists and of the First International Conference on Conservation of the Wolf*, ed. by Douglas H. Pimlott (Morges: IUCN, 1975), p. 12.
47. David W. Macdonald, Georgina M. Mace and Steve Rushton, 'British Mammals: Is There a Radical Future?', in *Priorities for the Conservation of Mammalian Diversity: Has the Panda Had Its Day?*, ed. by Abigail Entwistle and Nigel Dunstone (Cambridge: Cambridge University Press, 2000), pp. 175–205 (p. 175).
48. Göran Ericsson and Thomas A. Heberlein, 'Attitudes of Hunters, Locals, and the General Public in Sweden Now That the Wolves are Back', *Biological Conservation*, 111 (2003), 149–59.
49. IUCN/SSC, *Guidelines for Reintroductions and Other Conservation Translocations* (Gland, Switzerland: IUCN Species Survival Commission, 2013), p. 4.
50. John D. C. Linnell and Craig R. Jackson, 'Bringing Back Large Carnivores to Rewild Landscapes', in *Rewilding*, ed. by Nathalie Pettorelli, Sarah M. Durant, and Johan T. du Toit (Cambridge: Cambridge University Press, 2019), pp. 248–79 (p. 264).
51. Scotland: The Big Picture, *Hearts and Minds: An Assessment of the Social and Cultural Barriers to Rewilding in Scotland* (2022), p. 97.
52. Ibid. For the headlines and news articles quoted, see Karl Grafton, 'Dive Bombers: Kids "Living in Fear" of Vicious Seagulls Who Have "Taken Over" Quiet Scots Street', *Scottish Sun*, 2 July 2021 <https://www.thescottishsun.co.uk/news/scottish-news/7345449/seagull-attacks-linwood-greenhill-crescent/>; 'Seagulls: Denis O'Donovan Calls for Cull of "Vicious" Birds', *BBC News*, 20 July 2015 <https://www.bbc.co.uk/news/world-europe-33601013>; Emma James, 'Ant Invasion: Flying Ants "Invade" UK on Hottest Day as "Mass Outpourings" of Critters Blight Brits', *Sun*, 18 July 2021 <https://www.thesun.co.uk/news/15622766/flying-ant-day-2021-hottest-day-skies/>; and Arthi Nachiappan, 'Red Alert Over Henley as Hungry Kites Attack Town', *Times*, 25 May 2021 <https://www.thetimes.com/life-style/food-drink/article/red-alert-over-henley-as-hungry-kites-attack-town-59p5qh7hf>.
53. M. J. Goulding and T. J. Roper, 'Press Responses to the Presence of Free-living Wild Boar (*Sus scrofa*) in Southern England', *Mammal Review*, 32 (2002), 272–82 (p. 276).
54. Monbiot, *Feral*, pp. 106–7.
55. Linnell and Alleau, 'Predators That Kill Humans', p. 361, and Linnell et al., 'The Fear of Wolves', p. 38.
56. Linnell et al., 'The Fear of Wolves', p. 6.
57. Ibid., pp. 5 and 38, and Alejandro Martínez-Abraín et al., 'Do Apex Predators Need to Regulate Prey Populations to be a Right Conservation Target?', *Biological Conservation*, 261 (2021). In the UK alone lightning kills an average of two people per year, but across the entirety of Europe and Russia between 1950 and 2000, deaths attributable to wolves averaged around 0.34 a year; Ian Convery

et al., 'The Case for Wolves in the UK', in *The Wolf: Culture, Nature, Heritage*, ed. by Convery et al., pp. 295–314 (p. 300). Between 2002 and 2020, there were no fatalities across Europe; John D. C. Linnell, Ekaterina Kovtun, and Ive Rouart, 'Wolf Attacks on Humans: An Update for 2002–2020', Norwegian Institute for Nature Research (2021), pp. 16–17.

58. It is suggested that the 'furious' phrase of the disease displays particularly strongly in wolves, with single rabid wolves attacking numerous people and animals in short spaces of time. Wolves with rabies have been described as 'the most dangerous rabid animal[s] of all'; Linnell et al., 'The Fear of Wolves', p. 14.
59. Monbiot, *Feral*, p. 113.
60. Linnell, Kovtun, and Rouart, 'Wolf Attacks on Humans: An Update for 2002–2020', pp. 3 and 14–16.
61. Linnell et al., 'The Fear of Wolves', p. 36.
62. Mech, 'Alpha Status, Dominance, and Division of Labor in Wolf Packs', p. 1197, and Gow, *Hunt for the Shadow Wolf*, p. 93.
63. L. David Mech, 'Who's Afraid of the Big Bad Wolf? – Revisited', *International Wolf*, 8.1 (1998), pp. 8–11 (p. 9).
64. Linnell et al., 'The Fear of Wolves', p. 5.
65. Ibid., p. 36.
66. Ibid., p. 16.
67. Ibid., p. 23.
68. Ibid., p. 36.
69. Roy Dennis, 'Re-introduction of Birds and Mammals to the British Isles', *Biologist*, 50 (2003), 20–4 (p. 23).
70. M. Zanni et al., 'The Wolf and the City: Insights on Wolves' Conservation in the Anthropocene', *Animal Conservation*, 26 (2023), 766–80.
71. Adam F. Smith et al., 'Quiet Islands in a World of Fear: Wolves Seek Core Zones of Protected Areas to Escape Human Disturbance', *Biological Conservation*, 276 (2022), 1–11.
72. J. Karlsson, M. Eriksson, and O. Liberg, 'At What Distance Do Wolves Move Away from an Approaching Human?', *Canadian Journal of Zoology*, 85, 1193–7.
73. Linnell and Alleau, 'Predators That Kill Humans', p. 363, and Mark E. McNay, 'Wolf-Human Interactions in Alaska and Canada: A Review of the Case History', *Wildlife Society Bulletin (1973-2006)*, 30 (2002), 831–43 (p. 833).
74. Jacqueline Huber et al., 'Wolves Living in Proximity to Humans: Summary of a First Enquiry on Wolf Behaviour Near Humans in Europe', *KORA Bericht*, 76 (2016), pp. 14–15.
75. Linnell, Kovtun, and Rouart, 'Wolf Attacks on Humans: An Update for 2002–2020', p. 16.
76. Mech, 'Who's Afraid of the Big Bad Wolf?', p. 11.
77. Rick Lamplugh, 'Creating a World of Wolf Haters', *Oregon Wild*, 12 December 2013 <https://www.oregonwild.org/creating-world-wolf-haters> [accessed 16 April 2024].
78. Linnell et al., 'The Fear of Wolves', p. 36.
79. Mech, 'Who's Afraid of the Big Bad Wolf', p. 9.
80. Linnell et al., 'The Fear of Wolves', p. 5.
81. Wilson, 'Could We Live with Reintroduced Large Carnivores?', p. 224, and Monbiot, *Feral*, p. 114.
82. Vincenzo Gervasi et al., 'Ecological Correlates of Large Carnivore Depredation on Sheep in Europe', *Global Ecology and Conservation*, 30 (2021), 1–13.
83. Colwell, *Beak, Tooth and Claw*, p. 51.

84. David W. Macdonald et al., 'Conserving Large Mammals: Are They a Special Case?', in *Key Topics in Conservation Biology 2*, ed. by David W. Macdonald and Katherine J. Willis (Chichester: Wiley & Sons, 2013), pp. 277–312 (p. 283).
85. Marvin, 'Wolves in Sheep's (and Others') Clothing', pp. 70–2.
86. Juan Carlos Blanco and Kerstin Sundseth, 'The Situation of the Wolf (*Canis lupus*) in the European Union: An In-depth Analysis', report of the N2K Group for DG Environment, European Commission (2023), p. 9.
87. Christopher J. Sandom and David W. Macdonald, 'What Next? Rewilding as a Radical Future for the British Countryside', in *Wildlife Conservation on Farmland: Managing for Nature on Lowland Farms*, ed. by David W. Macdonald and Ruth E. Feber (Oxford: Oxford University Press, 2015), pp. 291–316 (p. 295).
88. Christopher J. Sandom and Sophie Wynne-Jones, 'Rewilding a Country: Britain as a Study Case', in *Rewilding*, ed. by Nathalie Pettorelli et al. (Cambridge: Cambridge University Press, 2019), pp. 222–47 (p. 228).
89. Sandom and Macdonald, 'What Next? Rewilding as a Radical Future for the British Countryside', p. 295.
90. Tom Williamson, 'Rewilding: A Landscape-history Perspective', *British Wildlife*, 33.6 (2022), pp. 423–9 (pp. 425–6).
91. Jonathan D. Gordon and Brennen Fagan, 'The First Farmers Often Made Landscapes More Biodiverse – Our Research Could Have Lessons for Rewilding Today', *The Conversation*, 26 July 2024 <https://theconversation.com/the-first-farmers-often-made-landscapes-more-biodiverse-our-research-could-have-lessons-for-rewilding-today-233272> [accessed 28 July 2024].
92. Coulthard, *A Short History of the World According to Sheep*, p. 67.
93. Monbiot, *Feral*, p. 157, and Robert A. Robinson and William J. Sutherland, 'Post-war Changes in Arable Farming and Biodiversity in Great Britain', *Journal of Applied Ecology*, 39 (2002), 157–76 (p. 158).
94. Numbers vary across the year – June surveys, which include lambs, are higher, while December surveys, which take place after lambs have been slaughtered, are lower. For the UK statistics for June and December 2024, see DEFRA, 'Livestock Populations in the United Kingdom at 1 June 2024', 27 March 2025 <https://www.gov.uk/government/statistics/livestock-populations-in-the-united-kingdom/livestock-populations-in-the-united-kingdom-at-1-june-2024>, and DEFRA, 'Livestock Populations in the United Kingdom at 1 December 2024', 27 March 2025 <https://www.gov.uk/government/statistics/livestock-populations-in-the-united-kingdom/livestock-populations-in-the-united-kingdom-at-1-december--2#:~:> [accessed 12 November 2025]. For the Ireland statistics for June and December 2024, see Ireland Central Statistics Office, 'Crops and Livestock Provisional June 2024', 09 October 2024 <https://www.cso.ie/en/releasesandpublications/ep/p-clsjp/cropsandlivestockprovisionaljune2024/>, and Ireland Central Statistics Office, 'Livestock Survey December 2024', 04 March 2025 <https://www.cso.ie/en/releasesandpublications/ep/p-lsd/livestocksurveydecember2024/> [accessed 12 November 2025]. For data on national sheep flock numbers around the world, see World Population Review, 'Sheep Population by Country 2025' <https://worldpopulationreview.com/country-rankings/sheep-population-by-country> [accessed 12 November 2025].
95. MacDonald, *Rebirding*, p. 150. As MacDonald notes, 13,260km^2 of Wales's 20,779km^2 area is given over to sheep pasture, which housed 9.6 million sheep in 2016, compared to the 3.1 million people who lived in Wales in the same year.
96. Sandom and Macdonald, 'What Next? Rewilding as a Radical Future for the British Countryside', p. 295.
97. Edwards and Kenyon, 'Wild Deer in Scotland', p. 23; T. H. Clutton-Brock, T. Coulson, and J. M. Milner, 'Red Deer Stocks in the Highlands of Scotland',

Nature, 429 (2004), 261–2 (p. 261); Gill, 'The Impact of Deer on Woodland Biodiversity', p. 3; and Monbiot, *Feral*, p. 159.

98. Robinson and Sutherland, 'Post-war Changes in Arable Farming and Biodiversity in Great Britain', p. 167.
99. Sandom et al., 'Exploring the Value of Wolves in Landscape-Scale Fenced Reserves', p. 249.
100. MacDonald, *Rebirding*, p. 153.
101. Much of the lamb eaten in Britain is imported from as far away as New Zealand; MacDonald, *Rebirding*, p. 151.
102. Sally Coulthard notes that 'soil degradation [...] could be reversed if we return to mixed farming systems which include plenty of pasture and nitrogen-fixing legumes grazed by sheep and cattle', while 'pastureland – especially if it is sown with long-rooted varieties of grass – also captures and locks away carbon'; *A Short History of the World According to Sheep*, pp. 276–7.
103. J. D. C. Linnell, *From Conflict to Coexistence? Insights from Multi-disciplinary Research into the Relationships Between People, Large Carnivores and Institutions* (Gland: Large Carnivore Initiative for Europe, 2013), p. 36.
104. Lee Schofield, 'The Three-Legged Stool: Wolves, Shepherds and Sheep', in *The Wolf: Culture, Nature, Heritage*, ed. by Convery et al., pp. 363–70 (pp. 369–70).
105. Urs Breitenmoser et al., 'Non-Lethal Techniques For Reducing Depredation', in *People and Wildlife, Conflict or Co-Existence?*, ed. by Rosie Woodroffe, Simon Thirgood, and Alan Rabinowitz (Cambridge: Cambridge University Press, 2005), pp. 49–71 (pp. 57–8).
106. John D. C. Linnell and Benjamin Cretois, 'Research for AGRI Committee – The Revival of Wolves and Other Large Predators and Its Impact on Farmers and their Livelihood in Rural Regions of Europe', European Parliament, Policy Department for Structural and Cohesion Policies, Brussels (2018), p. 10.
107. Ibid.
108. The practice is mentioned in Aristotle's (384–322 BCE) *History of Animals* (John D. C. Linnell and Nicolas Lescureux, 'Livestock Guarding Dogs: Cultural Heritage Icons with a New Relevance for Mitigating Conservation Conflicts', Norwegian Institute for Nature Research (2015), p. 11), though it is likely even older – it is thought that livestock protection dogs originated in ancient Mesopotamia after the development of sheep and goat farming, though they were not bred specifically for the purpose but were simply dogs raised alongside and bonded to livestock; Thomas M. Gehring, Kurt C. VerCauteren, and Jean-Marc Landry, 'Livestock Protection Dogs in the 21st Century: Is an Ancient Tool Relevant to Modern Conservation Challenges?', *USDA National Wildlife Research Center Staff Publications*, 919 (2010), 299–308 (p. 301).
109. Suzanne Asha Stone, *Livestock and Wolves: A Guide to Nonlethal Tools and Methods to Reduce Conflicts*, second edn (Washington, DC: Defenders of Wildlife, 2016), p. 6.
110. Ibid.; Gehring, VerCauteren, and Landry, 'Livestock Protection Dogs in the 21st Century', p. 303; Breitenmoser et al., 'Non-Lethal Techniques For Reducing Depredation', p. 63; and Cosmin Marius Ivaşcu and Alina Biro, 'Coexistence Through the Ages: The Role of Native Livestock Guardian Dogs and Traditional Ecological Knowledge as Key Resources in Conflict Mitigation between Pastoralists and Large Carnivores in the Romanian Carpathians', *Journal of Ethnobiology*, 40 (2020), 465–82 (p. 475).
111. Martin E. Smith et al., 'Review of Methods to Reduce Livestock Depredation: I. Guardian Animals', *Acta Agriculturae Scandinavica, Section A—Animal Science* 50 (2000), 279–90 (pp. 285–7).

112. Stone, *Livestock and Wolves*, pp. 11–12.
113. Andrew C. Kitchener, 'Re-wilding Ireland: Restoring Mammalian Diversity or Developing New Mammalian Communities?', *The Irish Naturalists' Journal*, All-Ireland Mammal Symposium 2009 (2012), pp. 4–13 (p. 11).
114. Carl R. Gustavson, 'An Evaluation of Taste Aversion Control of Wolf (*Canis lupus*) Predation in Northern Minnesota', *Applied Animal Ethology*, 9 (1982), 63–71, and Monbiot, *Feral*, p. 114.
115. Convery et al., 'The Case for Wolves in the UK', pp. 303–4.
116. Martin Drenthen, 'Finding Common Ground with Wolves: Interspecies Communication in a Shared Landscape', in *The Wolf: Culture, Nature, Heritage*, ed. by Convery et al., pp. 287–94 (pp. 292–3).
117. Wood River Wolf Project <https://www.woodriverwolfproject.org/history> [accessed 15 March 2024].
118. Wildlife Friendly Enterprise Network, 'Certified Wildlife Friendly® Enterprises that Protect Wolves' <https://www.wildlifefriendly.com/wolf> [accessed 13 November 2025].
119. Lily M. van Eeden et al., 'Public Willingness to Pay for Gray Wolf Conservation That Could Support a Rancher-led Wolf-livestock Coexistence Program', *Biological Conservation*, 260 (2021), 1–10.
120. Adrian Treves and Jeremy Bruskotter, 'Tolerance for Predatory Wildlife', *Science*, 344 (2014), 476–7 (p. 476).
121. Philip J. Nyhus et al., 'Bearing the Costs of Human-wildlife Conflict: The Challenges of Compensation Schemes', in *People and Wildlife, Conflict or Co-Existence?*, ed. by Woodroffe, Thirgood, and Rabinowitz, pp. 107–21 (p. 118), citing David Cope, Juliet Vickery, and Marcus Rowcliffe, 'From Conflict to Coexistence: A Case Study of Geese and Agriculture in Scotland', in *People and Wildlife*, ed. by Woodroffe, Thirgood, and Rabinowitz, pp. 176–91.
122. Francisco J. Santiago-Avila, Ari M. Cornman, and Adrian Treves, 'Killing Wolves to Prevent Predation on Livestock May Protect One Farm but Harm Neighbors', *PLoS ONE*, 13 (2018), 1–20 (p. 15), and Robert B. Wielgus and Kaylie A. Peebles, 'Effects of Wolf Mortality on Livestock Depredations', *PLoS ONE*, 9 (2014), 1–16 (pp. 10–12).
123. Colwell, *Beak, Tooth and Claw*, p. 190.
124. Scotland: The Big Picture, *Hearts and Minds*, pp. 14 and 18.
125. Marvin, 'Wolves in Sheep's (and Others') Clothing', p. 74, and Helene Figari and Ketil Skogen, 'Contemporary Public Images of the Wolf', in *The Wolf: Culture, Nature, Heritage*, ed. by Convery et al., pp. 145–50 (p. 147).
126. Crumley, *The Last Wolf*, p. 146.
127. Figari and Skogen, 'Contemporary Public Images of the Wolf', p. 147.
128. See p. 125 above.
129. Gwynn and Symeonakis, 'Rule-based Habitat Suitability Modelling', pp. 6 and 16.
130. <https://www.reddit.com/r/AskHistory/comments/1cd5276/did_wolves_actually_live_in_the_uk> [accessed 24 June 2024].
131. Monbiot, *Feral*, pp. 163–4.
132. Melissa Kite, 'The Missing Lynx? How the New Craze on the Fundamentalist Wing of the Naturalist Lobby Could Destroy the Countryside', *Spectator*, 24 September 2016 <https://www.spectator.co.uk/article/the-missing-lynx> [accessed 21 October 2020].
133. The phenomenon of forgetting what the landscape once was is known as 'shifting baseline syndrome'. Each generation assumes and accepts that the landscape they know is simply how it is and how it has always been, leading us all to preserve what is rather than what was, what could be, or what should be. As the

environment becomes ever more depleted, the 'baselines' to which we adhere fall lower and lower, to the point where it is degraded and damaged habitats that are protected and preserved.

134. Yalden, *History of British Mammals*, p. 270.
135. MacDonald et al., 'Conserving Large Mammals: Are They a Special Case?', p. 280.
136. Guillaume Chapron et al., 'Recovery of Large Carnivores in Europe's Modern Human-dominated Landscapes', *Science*, 346 (2014), 1517–19 (p. 1518).
137. Linnell et al., 'The Fear of Wolves', p. 7.
138. Jeremy T. Bruskotter et al., 'Removing Protections for Wolves and the Future of the U.S. Endangered Species Act (1973)', *Conservation Letters*, 7 (2014), 401–7 (p. 405).
139. Chapron et al., 'Recovery of Large Carnivores in Europe's Modern Human-dominated Landscapes', p. 1518.
140. Macdonald et al., 'Conserving Large Mammals: Are They a Special Case?', p. 293, and Chapron et al., 'Recovery of Large Carnivores in Europe's Modern Human-dominated Landscapes', p. 1518.
141. Ethan D. Doney et al., 'Wild About Wolves: Using Collaboration and Innovation to Bridge Parks, People, and Predators', *Conservation Science and Practice*, 5.7 (2023), 1–14.
142. Drenthen, 'Finding Common Ground with Wolves', p. 288, and Macdonald et al., 'Conserving Large Mammals: Are They a Special Case?', pp. 293–4.
143. While other potentially viable areas exist, research currently centres around the Highlands as the most suitable place for an initial reintroduction. Although deer numbers are increasing in England and Wales, neither is a strong contender – Wales has extensive upland sheep farming, while much of the landscape in England is densely populated by people.
144. Gwynn and Symeonakis, 'Rule-based Habitat Suitability Modelling'.
145. Wilson, 'Could We Live with Reintroduced Large Carnivores?', p. 213.
146. Convery et al., 'The Case for Wolves in the UK', p. 298.
147. MacDonald, *Rebirding*, p. 122. As MacDonald points out, Scotland's deer estates cover more than 18,000 square kilometres, an area twice as large as Yellowstone National Park and which is extremely sparsely populated by both people and livestock; *Rebirding*, p. 128.
148. Monbiot, *Feral*, p. 99.
149. Wilson, 'Could We Live with Reintroduced Large Carnivores?', p. 213. See also Gwynn and Symeonakis, 'Rule-based Habitat Suitability Modelling'.
150. Scottish Government, 'Sheep and Lamb Processing: Assessment' (2019) <https://www.gov.scot/publications/assessment-opportunities-retain-increase-sheep-lamb-processing-scotland/pages/3/> [accessed 17 April 2024].
151. Hickey, *Wolves in Ireland*, pp. 109–10.
152. Murphy et al., 'GIS-integrated Agent-based Simulations to Model Wolf Reintroduction Management Scenarios in Ireland'.
153. See Daniel S. Licht et al., 'Using Small Populations of Wolves for Ecosystem Restoration and Stewardship', *BioScience*, 60 (2010), 147–53.
154. Gwynn and Symeonakis, 'Rule-based Habitat Suitability Modelling', p. 15.
155. Ibid.
156. Murphy et al., 'GIS-integrated Agent-based Simulations to Model Wolf Reintroduction Management Scenarios in Ireland'.
157. Licht et al., 'Using Small Populations of Wolves for Ecosystem Restoration and Stewardship', p. 150.
158. Sandom et al., 'Exploring the Value of Wolves in Landscape-Scale Fenced Reserves', pp. 245 and 250.

159. Sandom et al., 'Exploring the Value of Wolves in Landscape-Scale Fenced Reserves', and Bull et al., 'Fences Can Support Restoration in Human-dominated Ecosystems'.
160. Manning, Gordon and Ripple, 'Restoring Landscapes of Fear with Wolves in the Scottish Highlands', p. 2318.
161. Sandom et al., 'Exploring the Value of Wolves in Landscape-Scale Fenced Reserves', p. 271.
162. William S. Lynn, '*Canis Lupus Cosmopolis*: Wolves in a Cosmopolitan World-view', *Worldviews*, 6 (2002), 300–27 (p. 319).
163. Sandom et al., 'Exploring the Value of Wolves in Landscape-Scale Fenced Reserves', p. 270.
164. Yalden, *History of British Mammals*, pp. 270–1; D. W. Yalden, 'Opportunities for Reintroducing British Mammals', *Mammal Review*, 16 (1986), 53–63 (pp. 57–60); and Macdonald, Mace, and Rushton, 'British Mammals: Is There a Radical Future?', p. 195.

Postscript: The Future of Wolfland

1. We are now in the midst of a period ominously known as the 'sixth mass extinction', or the 'Holocene extinction'. While the previous five mass extinctions (such as the extinction of the dinosaurs) that scientists have identified were the result of natural causes, humans are entirely to blame for this latest extinction event. Hundreds of thousands of species are at imminent risk of disappearing, while many have already been lost in the past five centuries which, if it were not for humans, would have taken 18,000 years to disappear naturally; Gerardo Ceballos and Paul R. Ehrlich, 'Mutilation of the Tree of Life Via Mass Extinction of Animal Genera', *Proceedings of the National Academy of Sciences*, 120.39 (2023), 1–6.
2. John Hannigan, *Environmental Sociology*, fourth edn (London: Routledge, 2023), p. 165.
3. Helen Phillips et al., 'The Biodiversity Intactness Index: Country, Region and Global-level Summaries for the Year 1970 to 2050 under Various Scenarios' [Data set], Natural History Museum (2021).
4. Hannigan, *Environmental Sociology*, p. 165.
5. Sigmund Freud, *Civilization and Its Discontents*, trans. by David McLintock (London: Penguin, 2002), p. 29.
6. Charles Foster, *Being a Beast* (London: Profile Books, 2016), p. 173.
7. 'The Somnambulists', in *Revolution and Other Essays* (New York: Macmillan, 1910), pp. 41–53 (p. 41).

Index

Page numbers in *italics* denote illustrations.